A CONCISE OVERVIEW OF THE FINITE ELEMENT METHOD

A CONCISE OVERVIEW OF THE FINITE ELEMENT METHOD

JOHN O. DOW

University of Colorado at Boulder

MOMENTUM PRESS
ENGINEERING

MOMENTUM PRESS, LLC, NEW YORK

A Concise Overview of the Finite Element Method

Copyright © Momentum Press®, LLC, 2015.

First published by Momentum Press®, LLC
222 East 46th Street, New York, NY 10017
www.momentumpress.net

ISBN-13: 978-1-60650-508-3 (print)
ISBN-13: 978-1-60650-509-0 (e-book)

Momentum Press Solid Mechanics Collection

Collection ISSN: 2375-8570 (print)
Collection ISSN: 2375-8589 (electronic)

Cover and interior design by Exeter Premedia Services Private Ltd., Chennai, India

10 9 8 7 6 5 4 3 2 1

Printed in the United States of America

ABSTRACT

This is the third book in a series designed to make the finite element method and adaptive refinement more accessible. The first book, *A Unified Approach to the Finite Element Method and Error Analysis Procedures*, laid the theoretical background for simplifying and extending the finite element and finite difference methods by introducing the use of a physically interpretable notation. This notation simplifies and extends computational mechanics by reversing the usual mathematical approach of abstracting the notation used in the development. By introducing variables that represent rigid body motions and strain quantities, the modeling capabilities of individual elements and the solution algorithms can be evaluated for accuracy by visual inspection. The second book, *The Essentials of Finite Element Modeling and Adaptive Refinement*, demonstrated the validity of the derivations contained in the first book and extended the simplification of element formulation and improved the error measures.

This book, as its title indicates, presents a straightforward introduction to the finite element method, error analysis, and adaptive refinement. As such, it provides an easy-to-read overview that allows the contents of other finite element books and finite element courses to be seen in perspective as the various procedures are encountered. Furthermore, it provides developments that improve the procedures contained in the standard finite element textbook. As a result, when this book is used alone or in conjunction with other presentations, the reader is capable of critically assessing the capabilities of the finite element method.

KEYWORDS

adaptive refinement, error analysis, finite element method, symbolic computation

CONTENTS

LIST OF FIGURES

LIST OF TABLES

PREFACE

When I first approached thermodynamics, I was lucky enough to find a cogent overview of the subject entitled *Understanding Thermodynamics* by H. C. Van Ness. This short (107-page) book provided a useful introductory explanation of the law of thermodynamics and their relationship to each other. The foreknowledge provided by this overview significantly improved my ability to learn the subject.

The primary objective of this book is to provide a clear and easy to read *overview* of the finite element method for undergraduates in engineering, applied mathematics, and physics. This book is not meant to compete with commonly used textbooks. It is designed to eliminate the *mystery* that often surrounds the study of a new discipline. That is to say, this book is designed to show the reader the purpose and function of the various components of the finite method and how they fit together.

The recommended background knowledge for making this book accessible is minimal. From calculus, readers should already be familiar with the *meaning* of a derivative and an integral. From linear algebra, they should have a basic *understanding* of the concept of linear independence. From engineering or physics, readers should be *familiar with* the concepts of equilibrium, Hooke's law, and potential energy.

Understanding the concepts presented in this book does not require an intimate knowledge of computer programming. However, Chapters 5 and 6 include MATLAB programs as appendixes. These programs are included for three reasons: (1) If any of the text seems ambiguous, the MATLAB programs with their extensive annotation will eliminate it because of the precision required in defining the computations; (2) the best way to learn programming skills is to dissect operating programs line-by-line; and (3) the capabilities presented provide the basis for expanding the material presented for class projects or publishable research.

Solid mechanics is chosen as the focus because of the importance of boundary conditions in the problems. Many introductory works address scalar problems where the treatment of boundary conditions is much simpler. Furthermore, most, if not all, scalar problems can be solved with the procedures for modeling solid mechanics problems.

The presentational style is driven by the point of view of two of my favorite technical authors. Cornelius Lanczos in his classic book *The Variational Principles in Mechanics* observes the following:

> Many of the scientific treatises of today are formulated in a half-mystical language, as though to impress the reader with the uncomfortable feeling that he is in the permanent presence of a superman. The present book is conceived in a humble spirit and written for humble people. (1966)

When Prof. O. C. Zienkiewicz presented the 1998 Timoshenko Medal lecture, he made the following comments:

> After contact with Timoshenko's *Theory of Elasticity*, I realized that even quite complex ideas could be presented in a lucid form. … I have tried to follow the master [Timoshenko] by avoiding the alternative process, very popular among some scientific writers. They follow the maxim …, which simply stated is "Why make it simple when you can make it complicated"

The following is a summary of the chapters of this overview of the finite element method.

Chapter 1—Introduction: The need to solve complex problems with approximate solution techniques is introduced. Then, the difficulties involved in producing solutions accurate enough for use are presented. Finally, procedures for achieving solutions of prescribed accuracy are outlined.

Chapter 2—Formulation of Global Stiffness Matrices: Many finite element books focus on the formation of the stiffness matrices before they clarify the role that elemental stiffness matrices play in the final model. This chapter starts with a stiffness matrix that is known to most readers from high school physics, namely, the force-displacement relationship of a linear spring. Then, the process of assembling these simple elements into the global stiffness matrix for a complete structure is presented in two ways. First, the basic theory of forming a potential energy expression and minimizing it in order to produce the equilibrium equations is presented.

Then, the element-by-element assembly process that is the shortcut used in all finite element programs is presented.

Chapter 3—Physically Interpretable Displacement Interpolation Functions: In solid mechanics problems, the primary unknowns are the displacements. However, the strains, which are functions of the derivatives of the displacements, are usually the quantities sought by the analyst.

In this chapter, the interpolation polynomials that approximate the displacement in the finite element method are expressed in terms of rigid body motions and strain quantities. In other words, the displacements are expressed in terms of the quantities that produce them and are important to the analyst. It is this strain-based notation that makes this book so accessible to readers new to the finite element method and provides avenues for new research. This notation makes the finite element method transparent to the reader in a way not available with the standard notation.

This physically interpretable notation achieves its usefulness because it reverses the usual approach common to mathematics. Instead of expressing equations in terms of a notation that has arbitrary meaning, this notation is directly related to the concepts of continuum mechanics. When quantities that are important to solid mechanics are embedded in the notation, the equations formed using this notation are directly related to the concepts and theory of continuum mechanics. As a result, the equations can be understood and evaluated by visual inspection.

Chapter 4—An Improved Stiffness Matrix Formulation Procedure: The use of the physically interpretable notation presented in Chapter 3 simplifies and improves the finite element stiffness matrix formulation procedure. The number of integrals that must be evaluated is reduced, and they are simplified. Furthermore, errors in the strain representation of the individual elements can be visually identified during the formulation process because they are expressed in terms of strain quantities.

As a result of these characteristics, the standard isoparametric element formulation procedure is rendered obsolete. The evaluation of fewer and simpler integrals eliminates the need for the approximate numerical integration procedure that is central to the isoparametric approach. In addition to complicating the formulation procedure, this numerical integration procedure introduces strain modeling errors into finite elements that do not have a *regular shape*. The very fact that strain modeling errors are inherent in any elements produced by the standard isoparametric process makes this process obsolete.

Chapter 5—The "distmesh_2d" Mesh Generation Program: The distmesh_2d mesh generation program is used to generate the finite element meshes for the Kirsch problems solved here. The mesh generator is the product of a PhD thesis by Per-Olof Persson. It is available online and is contained in annotated form in the appendix of this chapter. Persson and Strang give the following reasons for its creation:

> Our goal is to develop a mesh generator that can be described in a few dozen lines of MATLAB. ... our chief hope is that users will take this code as a starting point for their own work. ..., but it can be simple and effective and public. (2004)

The introduction and application of an operational mesh generation program in a book that provides an overview of the finite element method is unique. This capability is included for two reasons. First, the availability of this capacity allows the reader to solve relatively complex problems that would otherwise be beyond their scope without putting in a concerted effort. Second, the availability of this capability in conjunction with the physically interpretable notation presented in Chapter 3 allows students new to the field to pursue state-of-the-art research in error analysis and mesh refinement. This is the case because the notation provides insights that are not available without this notation.

Chapter 6—Formulation of Finite Element Model of the Kirsch Problem: This chapter integrates the concepts presented in the previous chapters by solving the Kirsch problem. This classic problem is often used to test the accuracy of approximate methods. It is chosen because of its simplicity and the existence of known stress concentrations.

The problem consists of a panel with a circular hole in the center that is loaded in tension. Two stress concentrations exist on the boundary of the interior cutout. In addition to integrating the previous material, this problem is used to demonstrate the error estimator and refinement guides developed and applied in the next two chapters.

Chapter 7—Pointwise Error Estimators: The need to identify the level of errors in the individual finite elements was outlined in the introduction and reinforced in the previous chapter. In this chapter, error estimators that evaluate the accuracy of individual points are presented. These error estimators are put on a solid theoretical basis. This is accomplished by showing that approximate finite difference solutions can be extracted from the finite element solutions. Since these two solutions must converge to the same result, any differences between the two approximate solutions are due to deficiencies in the finite element model. This error estimator identifies the level of error in the individual finite elements.

Chapter 8—Simple and Effective Refinement Guides: Procedures for identifying the number of refinements that must be given to individual finite elements in order to improve the model are developed. A simple and effective refinement guide is developed and presented. It is based on the level of error estimated by the process developed in the previous chapter and the prescribed level of acceptable error.

The refinement guide presented produces rapid convergence of the solution to the specified level of accuracy in only two or three iterations of the adaptive refinement process that is outlined in the introduction and applied in this chapter. As would be expected, the more closely the initial model represents the problem, the quicker the adaptive refinement process converges to an accurate solution. A more detailed approach for identifying the number of subdivisions that must be given to an element for rapid convergence, which was developed by the author in a previous book, is outlined.

Chapter 9—Summary and Observations: This chapter compactly summarizes the details of the previous eight chapters and outlines approaches for further study and opportunities for possible research.

CHAPTER 1

INTRODUCTION

1.1   PROBLEM DEFINITION

Exact solutions to real-world solid mechanics problems are generally impossible to find. The geometry, boundary conditions, and loading conditions are often too complex to be captured by exact solution techniques. Consequently, approximations solutions are sought.

The finite element method is widely used to approximate such difficult problems. In this procedure, the problem is subdivided into a finite number of simple geometric shapes known as *finite elements*. Then, the equilibrium equations are formed and solved for the displacements at the finite number of nodes that are in the model. Finally, the stresses and strains are extracted from the nodal displacements.

Figure 1.1 depicts the mesh for a finite element model of a square panel with a circular hole. In this case, the finite elements are three-node triangles with nodes at the vertices.

In a finite element model, the displacements on an individual element are approximated by interpolating the nodal displacements. The displacements u(x,y) and v(x,y) at any point on a three-node element are found using the following interpolation polynomials:

$$u(x, y) = N_1 u_1 + N_2 u_2 + N_3 u_3$$

$$(1.1)$$

$$v(x, y) = N_1 v_1 + N_2 v_2 + N_3 v_3$$

where the N's are linear functions in the form $a_1 + a_2 x + a_3 y$ and the u_i's and v_i's are the nodal displacements of the triangle.

Unfortunately, *a significant difficulty arises* when an approximate solution technique is used. Since the solution is an approximation, how can we know that an approximate solution is accurate enough to be useful?

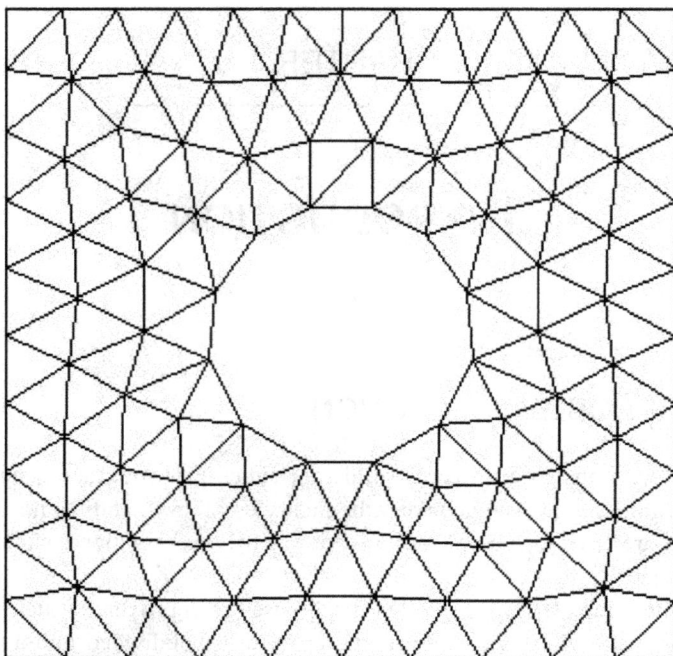

Figure 1.1. A finite element model.

We cannot compare it to the exact result because the exact solution cannot be found.

In other words, it appears that we have replaced an impossible problem with a problem that seems to be nearly as difficult. Instead of facing the impossibility of finding an exact solution, we face two formidable tasks. First, we must determine how to assess the accuracy of an approximate finite element solution. Then, we must devise ways to improve the finite element model so that the solution is accurate enough for use.

1.2 OBJECTIVES

The *overall objective* of this book is to provide a compact, intuitive, and theoretically solid presentation of a procedure for producing approximate solutions *with a predefined level of accuracy*. The presentation focuses on the finite element method and is aimed at readers who are new to computational mechanics. This objective is accomplished as follows:

1. The finite element method is introduced in a clear and concise manner.

2. Error estimation procedures that assess the accuracy of a solution are presented in an intuitive manner with a solid theoretical foundation.

3. A procedure is presented for rapidly improving the mesh with a rational, analytic method.

In combination, these three procedures define a capacity referred to as *adaptive refinement*.

1.3   AN OVERVIEW OF ADAPTIVE REFINEMENT

The adaptive refinement procedure is an iterative process that improves finite element models so that the solutions will possess a predetermined level of accuracy. This process is shown schematically in Figure 1.2.

As shown in the schematic diagram, the process begins by forming *an initial finite element model*. This model consists of the stiffness matrix, boundary conditions, and applied loads for the problem being solved. This problem is then solved for the nodal displacements. The stresses and strains used to evaluate the accuracy of the solution are extracted from the displacement approximations.

Once the solution portion of the procedure is completed, the *errors in each element are estimated*. Then, the accuracy of the solution is *evaluated*. If the error in *every element* falls below a predefined level, the analysis is stopped. If the error in any element exceeds the predefined error threshold, the adaptive refinement of the model proceeds.

If the model contains elements with excessive levels of error, it must be improved. This is accomplished by determining the number of subdivisions that must be given to each high-error element in order to achieve the desired level of accuracy. After the needed refinements are identified, the

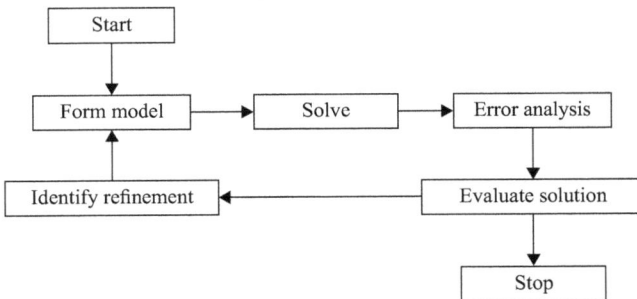

Figure 1.2. Adaptive refinement schematic.

subdivided elements are incorporated into a new finite element model. The process is repeated until the level of error in every element is acceptable.

1.4 AN EXAMPLE OF ADAPTIVE REFINEMENT

The result of the application of the adaptive refinement process is presented in this section. The overall geometry of the problem is shown in Figure 1.1. The panel is put in tension by stretching it with the application of equal and opposite distributed loads on the left and right ends of the panel. Stress concentrations exist at the top and bottom of the internal cutout.

Only one-quarter of the panel needs to be modeled because of symmetries that exist in this problem. As a result of these symmetries, certain displacements in the problem are known to be zero. These zero displacements are shown on the left-hand edge and the bottom edge of the problem with arrows in Figure 1.3.

The displacement in the x direction on the left side of the problem and the displacement in the y direction on the bottom edge of the problem are known to be zero. The nodal loads that represent the uniformly distributed loads on the right-hand edge of the panel are denoted by arrow heads.

The meshes in Figure 1.3 differ from the mesh in Figure 1.1. The mesh in Figure 1.1 consists of uniform triangles. The meshes in Figure 1.3 are graduated. The sizes of the individual finite elements decrease as they get closer to the circular internal cutout. This occurs because at some point in the adaptive refinement process, the errors in these elements exceeded

(a)   (b)

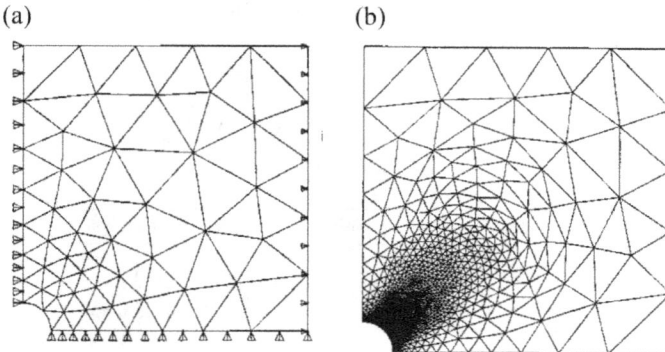

Figure 1.3. Adaptively refined stress concentration (Kirsch problem): (a) initial mesh and (b) adaptively refined mesh.

the prescribed level of acceptable error. Since the stresses change rapidly in the neighborhood of a stress concentration, the smaller elements capture the changes more accurately than do larger elements.

When the problem is analyzed with the initial mesh shown as Figure 1.3a, the error at the stress concentration exceeds 70 percent. After the model is refined as shown in Figure 1.3b, the error is below 5 percent (Dow 1999, 415). The first mesh contains 215 nodes and the second mesh contains 5,700 nodes.

1.5   SPECIAL FEATURES

As mentioned earlier, the displacements in the finite element model are approximated in the individual elements with interpolation polynomials. In the developments presented in Chapter 3, the arbitrary coefficients, the a's and b's that exist in the interpolation polynomials, are replaced with coefficients that have significant physical meaning in solid mechanics.

Specifically, the a and b coefficients are replaced with strains quantities (see Figure 3.1 and Equation 3.4). This change in notation produces the following important advantages:

1. The formulation of element stiffness matrices that form the equilibrium equations is simplified.
2. A solid theoretical basis is given to the error estimators.
3. The procedures for identifying the number of subdivisions to give to an element with an excessive level of error are put on a rational foundation.

As we will see in the main text, this change in notation allows new approaches to error analysis and mesh refinement strategies to be developed. As a result, new research opportunities are made available to both new and experienced finite element practitioners.

This universal opportunity exists because the physically based notation provides a fresh starting point for developing these capabilities. Everyone is close to the starting line when procedures are reformulated with a notation that provides a clear vision of the problems.

CHAPTER 2

Formulation of Global Stiffness Matrices

2.1 INTRODUCTION

The finite element method is a direct extension of the matrix analysis of trusses (Turner et al, 1956).* The extension to two- and three-dimensional problems produced a profound change in computational mechanics. As a consequence of this change, higher dimension problems with complex boundary and loading conditions that were previously unapproachable could now be approximated.

The extension to multidimensional problems is due to changes in the interpolation functions. Finite element stiffness matrices are formed with two- and three-displacement interpolation functions instead of the one-dimensional function that is used for truss elements (see Figure 3.1 and Equation 3.4). In both cases, these interpolation functions approximate the displacements in terms of the nodal displacements.

2.2 OBJECTIVE

Since the finite element method is an extension of the analysis of skeletal structures, computational similarities exist between the two techniques. In both cases, the stiffness matrices for individual elements are assembled to form a global stiffness matrix. In this chapter, we exploit the computational similarities by using truss examples to present processes for forming global stiffness matrices that are also applicable for finite element models.

---

*Turner's work is one of the first finite element papers.

The objective of this chapter is to present both the theoretical and practical procedures for assembling global stiffness matrices. Trusses are featured because the matrices involved with one-dimensional problems are relatively small. This makes the presentation of the assembly process more compact and easier to follow.

The global stiffness matrices are formed for both trusses and finite element models by forcing the displacements of the nodes of individual elements with common locations to have the same displacements. That is to say, displacement compatibility is imposed at the nodes of the individual elements.

In order to reinforce the similarity between the two analysis techniques and to justify the use of truss examples in this development, let us interpret the structure shown in Figure 2.1 as both a skeletal structure and a finite element model.

If the lines connecting any two nodes are considered as bars, the structure shown can be interpreted as a plane truss. In other words, the structure is a truss if the elements connecting nodes 1 and 2 and, say, nodes 4 and 5 are bar elements. From this point of view, the truss consists of 12 bar elements and 7 nodes.

There are displacements in the x and y direction at each of the 7 nodes for a total of 14 degrees of freedom, for example, u_5 and v_5 at node 5. The global stiffness matrix is formed by forcing each of the bars that connect at the same node to have identical displacements. For example, the three bars that connect at node 5 are forced to have the same displacement as node 5.

If, instead, each of the six triangular regions contained in Figure 2.1 is thought of as a three-node finite element, the figure can be considered

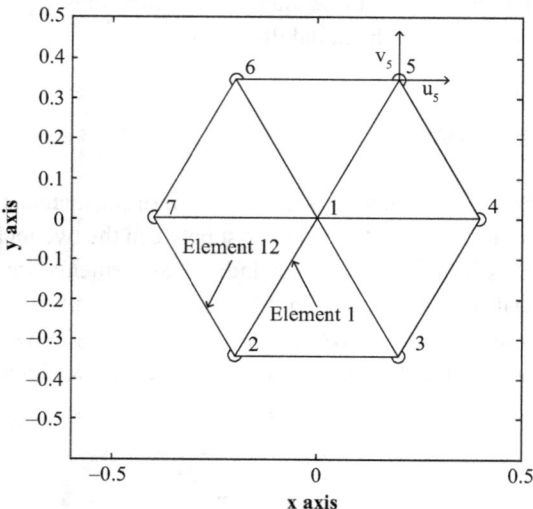

Figure 2.1. A generic structure.

to be a plane stress model. In this interpretation, each of the three-node elements contains the central node and two nodes on the boundary of the figure, for example, nodes 1, 4, and 5.

As was the case for the truss model, there is a u and v displacement at each of the seven nodes. Therefore, the finite element model contains the same 14 degrees of freedom as the truss model. The global stiffness matrix for a finite element model is formed by requiring the nodes of the individual elements that meet at a common point to exhibit the same displacement. For example, the nodes of the two, three-node elements that connect at node 5 will be forced to have the same displacement as node 5.

2.3   APPROACH

The process for creating a global stiffness matrix is a five-step process. Each of the five steps in the outline that follows is keyed to the equations that develop the global stiffness matrix in the next section.

A matrix equation representing the strain energy for each of the individual elements of the structure is formed in *its local coordinate system* (step 1 is accomplished in Equation 2.1). These individual strain energy matrices expressed in local coordinates are then inserted into a large matrix (step 2 is accomplished in Equation 2.2). The strain energy expression formed in step 2 must be transformed from local to global coordinates in order to produce the global strain energy expression. The coordinate transformation from local to global coordinates is shown in Equation 2.3. The strain energy expression is transformed from local to global coordinates in Equation 2.4.

A work function expressed in terms of equivalent nodal loads and the nodal displacements is formed. Then, the strain energy and the work function expressed in terms of the global coordinates are summed to produce the potential energy expression that provides the basis for the mathematical model (this portion of step 5 is accomplished in Equation 2.5).

The final mathematical model, which consists of a global stiffness matrix and an applied load vector, is produced when the principle of minimum potential energy is applied to the potential energy function (see Equation 2.6). The global stiffness matrix emerges from the strain energy portion of the potential energy function (see Equation 2.7).

2.4   THE FORMULATION OF THE FINITE ELEMENT STRUCTURAL MODEL

This section develops the procedure for creating the mathematical representation of a truss or a finite element model. As we will see, the direct formulation of the structural model is straightforward but the process is not computationally efficient. The matrices involved are extremely large and sparse. In practice, the model is formed with a less obvious approach that is computationally efficient (McGuire, Gallagher, and Ziemian 2000). Both procedures for forming the global stiffness matrices are presented in this chapter.

The unwieldy but theoretically transparent development of the global stiffness matrix consists of the following five steps:

Step 1: The strain energy expression for each of the individual elements is generated and has the following form:

$$SE_n = 1/2\{d\}_n^T [K]_n \{d\}_n \qquad (2.1)$$

where $\{d\}_n$ is a column matrix that contains the displacement degrees of freedom of the nth element in local coordinates and $[K]_n$ is the symmetric stiffness matrix that describes the force-displacement properties of the nth element.

Step 2: The elemental strain energy expressions are assembled into one large matrix equation to give the following:

$$SE_{Total} = 1/2 \left\{ \begin{array}{c} \{d\}_1^T \\ \{d\}_2^T \\ \vdots \\ \{d\}_n^T \end{array} \right\}^T \left[\begin{array}{cccc} [K]_1 & 0 & 0 & 0 \\ 0 & [K]_2 & 0 & 0 \\ 0 & 0 & \ddots & 0 \\ 0 & 0 & 0 & [K]_n \end{array} \right] \left\{ \begin{array}{c} \{d\}_1 \\ \{d\}_2 \\ \vdots \\ \{d\}_n \end{array} \right\} \qquad (2.2)$$

Step 3: The transformation from local to global coordinates is formed and has the following structure:

$$\left\{ \begin{array}{c} \{d\}_1 \\ \{d\}_2 \\ \vdots \\ \{d\}_n \end{array} \right\} = \left[\begin{array}{c} [T]_1 \\ [T]_2 \\ \vdots \\ [T]_n \end{array} \right] \{u\}_{Global} \qquad (2.3)$$

where $[T]_n$ is the $n_{elemental} \times n_{Global}$ transformation matrix that takes the local coordinates of the nth element to the global coordinate system and $\{u\}_{Global}$ is the column vector of the global coordinates.

Step 4: The global strain energy, that is, the total strain energy expressed in global coordinates, is formed by substituting Equation 2.3 into Equation 2.2 to give the following:

$$SE_{Global} = 1/2\{u\}_{Global}^T \begin{bmatrix} [T]_1 \\ [T]_2 \\ \vdots \\ [T]_n \end{bmatrix}^T \begin{bmatrix} [K]_1 & 0 & 0 & 0 \\ 0 & [K]_2 & 0 & 0 \\ 0 & 0 & \ddots & 0 \\ 0 & 0 & 0 & [K]_n \end{bmatrix} \begin{bmatrix} [T]_1 \\ [T]_2 \\ \vdots \\ [T]_n \end{bmatrix} \{u\}_{Global}$$

(2.4a)

When the kernel of Equation 2.4a is expanded, the equation takes on the following form:

$$SE_{Global} = 1/2\{u\}_{Global}^T \begin{bmatrix} [T]_1^T [K_1][T]_1 \\ +[T]_2^T [K_2][T]_2 + \cdots \\ +[T]_n^T [K_n][T]_n \end{bmatrix} \{u\}_{Global}$$

(2.4b)

$$SE_{Global} = 1/2\{u\}_{Global}^T [K]_{Global} \{u\}_{Global}$$

(2.4c)

Although, Equation 2.4 is theoretically correct and relatively easy to understand, it is not used in practice because of the size of the problems. It is not uncommon for practical problems to have several thousand or several hundred thousand elements and a corresponding number of degrees of freedom. As a consequence, each of the transformation matrices $[T]_i$ in Equation 2.4b can be very large and very sparse making the computations impractical.

Equation 2.4b provides the basis for a computationally efficient procedure for forming global stiffness matrices that is used in practice. This procedure will be demonstrated in a later section.

Step 5: The governing equations for a structures problem are formed when the principle of minimum potential energy is applied. The potential energy function is formed when a work function is added to the strain energy expression defined in the previous step to give the following:

$$PE = SE_{Global} + W$$

(2.5a)

$$PE = 1/2\{u\}_{Global}^T [K]_{Global} \{u\}_{Global} + \{u\}_{Global}^T \{F\}_{Global}$$

(2.5b)

where $\{F\}_{Global}$ is the vector of nodal loads applied to the structure.

The application of the principle of minimum potential energy produces the following result for each of the n degrees of freedom contained in the model:

$$\frac{\partial PE}{\partial u_i} = 1/2 \begin{Bmatrix} 0 \\ 0 \\ \vdots \\ 1 \\ \vdots \\ 0 \\ 0 \end{Bmatrix}^T \left[K \right]_{Global} \{u\}_{Global} + 1/2 \{u\}_{Global}^T \left[K \right]_{Global} \begin{Bmatrix} 0 \\ 0 \\ \vdots \\ 1 \\ \vdots \\ 0 \\ 0 \end{Bmatrix} + \begin{Bmatrix} 0 \\ 0 \\ \vdots \\ 1 \\ \vdots \\ 0 \\ 0 \end{Bmatrix}^T$$

$$\{F\}_{Global} = 0 \tag{2.6}$$

When the symmetry of the global stiffness matrix is invoked and the n equations are consolidated, the mathematical model for the structure becomes the following:

$$\left[K \right]_{Global} \{u\}_{Global} + \{F\}_{Global} = 0 \tag{2.7}$$

where $[K]_{Global}$ is an n × n matrix and $\{u\}_{Global}$ and $\{F\}_{Global}$ are n × 1 column vectors.

Equation 2.7 defines the structure of the finite element model for a problem. When Equation 2.7 is compared to Equation 2.5b, we can see that the finite element model can be formed without actually applying the principle of minimum potential energy. The governing equations can be formed directly by computing the global stiffness matrix, forming the applied load vector and relating them as indicated by Equation 2.7. The need to explicitly apply the principle of minimum potential energy is eliminated.

2.5 A DEMONSTRATION OF GLOBAL STIFFNESS MATRIX FORMULATION

One component of the kernel of the strain energy expression contained in Equation 2.4b will be developed in this section. On the one hand, this is done to demonstrate the process. On the other hand, the result provides a standard against which to compare the result produced by the computationally efficient approach for forming global stiffness matrices presented in the next section.

This demonstration consists of computing the contribution of element 12 of the truss shown in Figure 2.1 to the global stiffness matrix. This computation consists of performing the triple matrix product contained in Equation 2.4b that is associated with element 12.

Element 12 connects nodes 2 to 7 of the truss. The local coordinates for the element are shown in Figure 2.2a. The global coordinates are shown in Figure 2.2b. This figure identifies the correspondence between the local and global coordinates that is used to form the transformation from local to global coordinates presented in Equation 2.8b.

The transformation matrix that relates the four local coordinates of element 12 to the 14 global coordinates of the overall problem is as follows:

$$\{d\}_{12} = [T]_{12}\{u\}_{Global} \tag{2.8a}$$

$$\begin{Bmatrix} u_1^{12} \\ u_2^{12} \\ v_1^{12} \\ v_2^{12} \end{Bmatrix} = \begin{bmatrix} 0 & 1 & 0 & 0 & 0 & 0 & 0 & 0 & 0 & 0 & 0 & 0 & 0 & 0 \\ 0 & 0 & 0 & 0 & 0 & 0 & 1 & 0 & 0 & 0 & 0 & 0 & 0 & 0 \\ 0 & 0 & 0 & 0 & 0 & 0 & 0 & 0 & 1 & 0 & 0 & 0 & 0 & 0 \\ 0 & 0 & 0 & 0 & 0 & 0 & 0 & 0 & 0 & 0 & 0 & 0 & 0 & 1 \end{bmatrix} \begin{Bmatrix} u_1 \\ u_2 \\ \vdots \\ u_{14} \\ v_1 \\ v_2 \\ \vdots \\ v_{14} \end{Bmatrix}_{Global} \tag{2.8b}$$

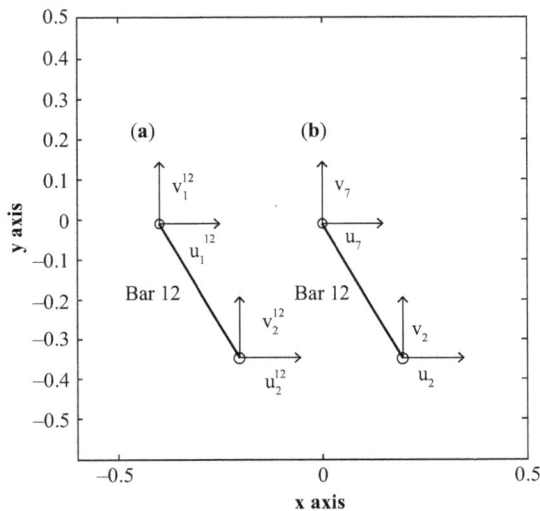

Figure 2.2. Coordinate definitions for element 12:
(a) element coordinates and (b) global coordinates.

The local stiffness matrix for Element 12 is given in generic form as follows:

$$[K]_{12} = \begin{bmatrix} k_{11}^{12} & k_{12}^{12} & k_{13}^{12} & k_{14}^{12} \\ k_{12}^{12} & k_{22}^{12} & k_{23}^{12} & k_{24}^{12} \\ k_{13}^{12} & k_{23}^{12} & k_{33}^{12} & k_{34}^{12} \\ k_{14}^{12} & k_{24}^{12} & k_{34}^{12} & k_{44}^{12} \end{bmatrix}$$

(2.9)

Note that this matrix is symmetric, that is, $k_{ij}^{12} = k_{ji}^{12}$.

When the triple matrix product for element 12, as identified in Equation 2.4b, is formed, the result is as follows:

$$[T]_{12}^{T}[K]_{12}[T]_{12} =$$

	u_1	u_2	u_3	u_4	u_5	u_6	u_7	v_1	v_2	v_3	v_4	v_5	v_6	v_7
u_1	0	0	0	0	0	0	0	0	0	0	0	0	0	0
u_2	0	k_{22}^{12}	0	0	0	0	k_{12}^{12}	0	k_{24}^{12}	0	0	0	0	k_{23}^{12}
u_3	0	0	0	0	0	0	0	0	0	0	0	0	0	0
u_4	0	0	0	0	0	0	0	0	0	0	0	0	0	0
u_5	0	0	0	0	0	0	0	0	0	0	0	0	0	0
u_6	0	0	0	0	0	0	0	0	0	0	0	0	0	0
u_7	0	k_{12}^{12}	0	0	0	0	k_{11}^{12}	0	k_{14}^{12}	0	0	0	0	k_{13}^{12}
v_1	0	0	0	0	0	0	0	0	0	0	0	0	0	0
v_2	0	k_{24}^{12}	0	0	0	0	k_{14}^{12}	0	k_{44}^{12}	0	0	0	0	k_{34}^{12}
v_3	0	0	0	0	0	0	0	0	0	0	0	0	0	0
v_4	0	0	0	0	0	0	0	0	0	0	0	0	0	0
v_5	0	0	0	0	0	0	0	0	0	0	0	0	0	0
v_6	0	0	0	0	0	0	0	0	0	0	0	0	0	0
v_7	0	k_{23}^{12}	0	0	0	0	k_{13}^{12}	0	k_{34}^{12}	0	0	0	0	k_{33}^{12}

(2.10)

The global coordinates corresponding to the rows and columns of the global stiffness matrix are listed across the top and along the left edge of the matrix. The identification of coordinates associated with the rows and columns of the partially complete global stiffness matrix lets us see how the elemental stiffness properties are distributed in the global stiffness matrix.

As can be seen in Equation 2.10, the stiffness properties of element 12 are associated with the coordinates that are present at nodes 2 and 7, namely, u_2, v_2, u_7, and v_7. This result will be used to validate the implicit approach for forming the stiffness matrix that is presented in the next section.

2.6 THE PRACTICAL FORMULATION OF THE GLOBAL STIFFNESS MATRIX

We will now demonstrate a procedure for locating the stiffness properties of an elemental stiffness matrix in the global stiffness matrix without actually performing the multiplications indicated in Equation 2.4b. This computationally efficient approach will be demonstrated for the case of element 12 used in the previous section.

The locations in the global stiffness matrix of the stiffness components of the individual stiffness matrices are contained in the data that defines a finite element model. The global coordinate associated with each of the elemental degrees of freedom is identified for each element in the input data.

The relationship between the local and global coordinates for element 12 is shown in Figure 2.2. When the two coordinate systems are compared, the following relationship exists between the local and the global coordinate systems:

$$
\begin{array}{lcccc}
\text{Local Coord. No.} & 1 & 2 & 3 & 4 \\
\text{Local Designation} & u_1^{12} & u_2^{12} & v_1^{12} & v_2^{12} \\
\text{Global Designation} & u_7 & u_2 & v_7 & v_2 \\
\text{Global Coord. No.} & 7 & 2 & 14 & 9
\end{array} \tag{2.11}
$$

The locations of the elements of the local stiffness matrix for element 12 in the global stiffness matrix are identified with the artifact shown in Equation 2.12. In this equation, the global coordinate numbers associated with the local coordinates are superimposed on the rows and columns of the elemental stiffness matrix as shown in the following:

$$
\begin{array}{c}
\begin{array}{cccc} 7 & 2 & 14 & 9 \end{array} \\
\begin{array}{c} 7 \\ 2 \\ 14 \\ 9 \end{array}
\begin{bmatrix}
k_{11}^{12} & k_{12}^{12} & k_{13}^{12} & k_{14}^{12} \\
k_{12}^{12} & k_{22}^{12} & k_{23}^{12} & k_{24}^{12} \\
k_{13}^{12} & k_{23}^{12} & k_{33}^{12} & k_{34}^{12} \\
k_{14}^{12} & k_{24}^{12} & k_{34}^{12} & k_{44}^{12}
\end{bmatrix}
\end{array} \tag{2.12}
$$

The row and column location in the global stiffness matrix for a stiffness element is identified by the global coordinate numbers associated with an individual k in Equation 2.12. For example, diagonal element k_{11}^{12} is located in row 7 and column 7 of the global stiffness matrix. Similarly, the two symmetric off-diagonal terms k_{34}^{12} are located in row 14 of column 9 and row 9 of column 14, respectively.

The fact that this implicit process locates the elements correctly can be seen by comparing the locations identified for each element in Equation 2.12 with the elements located in Equation 2.10, which was formed by multiplying out the strain energy expression. Note that the off-diagonal elements are located symmetrically in the global stiffness matrix.

The process of locating an element stiffness matrix in the global stiffness matrix will be demonstrated with another example. The objective of this example is to show how elements that are connected to each other in the structure are related in the global stiffness matrix.

The implicit coordinate transformation for element 1 in Figure 2.1 that is similar to Equation 2.11 is the following:

$$
\begin{array}{lcccc}
\text{Local Coord. No.} & 1 & 2 & 3 & 4 \\
\text{Local Designation} & u_1^1 & u_2^1 & v_1^1 & v_2^1 \\
\text{Global Designation} & u_2 & u_1 & v_2 & v_1 \\
\text{Global Coord. No.} & 2 & 1 & 9 & 8
\end{array}
\tag{2.13}
$$

When the global coordinate numbers are associated with the local stiffness matrix for element 1, we have the following:

$$
\begin{array}{c}
\begin{array}{cccc} 2 & 1 & 9 & 8 \end{array} \\
\begin{array}{c} 2 \\ 1 \\ 9 \\ 8 \end{array}
\begin{bmatrix}
k_{11}^1 & k_{12}^1 & k_{13}^1 & k_{14}^1 \\
k_{12}^1 & k_{22}^1 & k_{23}^1 & k_{24}^1 \\
k_{13}^1 & k_{23}^1 & k_{33}^1 & k_{34}^1 \\
k_{14}^1 & k_{24}^1 & k_{34}^1 & k_{44}^1
\end{bmatrix}
\end{array}
\tag{2.14}
$$

These elemental stiffness quantities can be added to the global stiffness matrix that already contains the stiffness properties for element 12 to give the following:

	1	2	3	4	5	6	7	8	9	10	11	12	13	14
1	k_{22}^1	k_{12}^1	0	0	0	0	0	k_{24}^1	k_{23}^1	0	0	0	0	0
2	k_{12}^1	$k_{22}^{12}+k_{11}^1$	0	0	0	0	k_{12}^{12}	k_{14}^1	$k_{24}^{12}+k_{13}^1$	0	0	0	0	k_{23}^{12}
3	0	0	0	0	0	0	0	0	0	0	0	0	0	0
4	0	0	0	0	0	0	0	0	0	0	0	0	0	0
5	0	0	0	0	0	0	0	0	0	0	0	0	0	0
6	0	0	0	0	0	0	0	0	0	0	0	0	0	0
7	0	k_{12}^{12}	0	0	0	0	k_{11}^{12}	0	k_{14}^{12}	0	0	0	0	k_{13}^{12}
8	k_{24}^1	k_{14}^1	0	0	0	0	0	k_{44}^1	k_{34}^1	0	0	0	0	0
9	k_{23}^1	$k_{24}^{12}+k_{13}^1$	0	0	0	0	k_{14}^{12}	k_{34}^1	$k_{44}^{12}+k_{33}^1$	0	0	0	0	k_{34}^{12}
10	0	0	0	0	0	0	0	0	0	0	0	0	0	0
11	0	0	0	0	0	0	0	0	0	0	0	0	0	0
12	0	0	0	0	0	0	0	0	0	0	0	0	0	0
13	0	0	0	0	0	0	0	0	0	0	0	0	0	0
14	0	k_{23}^{12}	0	0	0	0	k_{13}^{12}	0	k_{34}^{12}	0	0	0	0	k_{33}^{12}

$$(2.15)$$

This partially constructed global stiffness matrix contains a feature that can be explained in physical terms. As shown in Figure 2.1, the two bar elements 1 and 12 are connected at node 2 in the assembled structure. The displacement in the x direction in the global system is identified as u_2, and this displacement corresponds to global coordinate 2. Similarly, the displacement at node 2 in the y direction is v_2 with the global coordinate number of 9.

As can be seen in Equation 2.15, there are stiffness properties for both elements 1 and 12 in the diagonal terms (2, 2) and (9, 9) and in the off-diagonal terms (2, 9) and (9, 2). The fact that these terms contain sums of stiffness elements from the two bars indicates that they have the same displacements and the stiffness of the overall structure is a combination of the stiffness of the two bars.

The locations of the other elements can be checked by comparing the locations identified by the artifact presented in Equation 2.12 with Equation 2.10. When this process is applied to all of the elements, the final global stiffness matrix is produced.

2.7   SUMMARY

The primary objective of this chapter has been accomplished. The theoretical and practical approaches for forming the global stiffness matrices for finite element models have been made available to readers new to computational mechanics. The assembly process is covered more extensively in matrix structural analysis book by McGuire, Gallagher, and Ziemian (2000) and in introductory finite element books.

It should be noted that the hexagonal truss featured in this chapter is used in the next chapter as vehicle for introducing the physically interpretable notation that provides the basis for the developments presented in later chapters. This notation can be viewed as a straightforward modification of the existing interpolation polynomials used in the finite element method.

In this modification, the coefficients are expressed in terms of strain quantities. Since strains are primary quantities that are sought in continuum mechanics, the resulting equations are directly related to continuum mechanics concepts. The use of this physically interpretable notation allows the equations to be interpreted visually. This capability is demonstrated in the next chapter.

The transparence provided by this modified notation is what makes this book useful to such a wide readership. On the one hand, the finite element method, including the components of the adaptive refinement process, is made more accessible to readers new to computational mechanics. On the other hand, the insight provided by the transparency of the notation lets advanced practitioners extend and improve the capabilities of computational mechanics.

The hexagonal truss structure shown in Figure 2.1 is a two-dimensional version of Buckminster Fuller's geodesic dome. The physically interpretable notation developed in the next chapter will be used to show that this truss has the same stiffness properties in every direction. This may seem counterintuitive for a structure composed of discrete elements, but it can be useful in practice when the directions of the critical loads are unknown. Structures composed of such trusses were tested by the author in 1991 for use in aerospace applications.

2.8   EXERCISES

1. Relate the local coordinates for the element connecting nodes 2 and 3 in Figure 2.1 to the global coordinates. For compactness, refer to this element as element 6.

2. Form a figure similar to Equation 2.14 and add the stiffness matrix for element 6 to the global stiffness matrix given by Equation 2.15.

3. Form the transformation matrix between the elemental and global coordinates for element 6 in Figure 2.1 that is similar to Equation 2.8.

4. Perform the triple matrix product given by Equation 2.4c and compare the result to the answer to Exercise 2.

5. Form the total stiffness matrix for the structure composed of three bar elements with the same stiffness characteristics that are in a straight line along the x axis. The stiffness matrix for each element is: $K = \begin{bmatrix} k & -k \\ -k & k \end{bmatrix}$. The capability of constructing stiffness matrices for simple structures like this is useful in testing algorithms, for example, eigenvalue solutions.

6. Add the stiffness characteristics of a single bar that connects nodes 1 and 3 to the global stiffness matrix formed in Exercise 5. This bar has half of the stiffness of the other bars.

CHAPTER 3

PHYSICALLY INTERPRETABLE DISPLACEMENT INTERPOLATION FUNCTIONS

3.1   INTRODUCTION

When the one-dimensional displacement interpolation polynomials of matrix structural analysis were extended to two dimensions, the finite element method was created. This extension has produced profound improvements to computational mechanics. The advent of the finite element method allows problems with complex boundary and loading conditions that were previously unapproachable to be successfully approximated (Turner et al. 1956).

The interpolation polynomials are specialized for specific application to solid mechanics problems in this chapter. This specialization is accomplished by replacing the arbitrary coefficients in the interpolation functions with coefficients that have physical meaning with respect to continuum mechanics. As we will see, this improved notation simplifies element formulation, identifies strain modeling errors during the formulation process, leads to a solid theoretical foundation for the error estimators, and allows the refinement guides to be directly related to the modeling capabilities of the element used to form the finite element model.

An example of this improvement is shown in Figure 3.1. In this figure, the standard and the improved forms of the displacement interpolation functions for a three-node triangle are contrasted.

As can be seen in Figure 3.1a, the coefficients of the standard form of the interpolation polynomials are free of content. These arbitrary coefficients are equally opaque to any problem to which they are applied. The meaning of these coefficients must be inferred by the context in which they are used. For example, they provide no direct assistance to the

(a)

$$u(x,y) = a_1 + a_2 x + a_3 y$$
$$v(x,y) = b_1 + b_2 x + b_3 y$$

(b)

$$u(x,y) = (u_{rb})_0 + (\varepsilon_x)_0 x + (\gamma_{xy}/2 - r_{rb})_0 y$$
$$v(x,y) = (v_{rb})_0 + (\gamma_{xy}/2 + r_{rb})_0 x + (\varepsilon_y)_0 y$$

Figure 3.1. Three-node element interpolation polynomials: (a) standard interpolation polynomials and
(b) physically interpretable interpolation polynomials.

understanding of the finite element method, its capabilities, or the results
of the analysis.

In contrast, the coefficients of the interpolation polynomials shown in
Figure 3.1b are expressed in terms of quantities that are directly related to
solid mechanics, namely, rigid body motions and strain quantities. These
quantities represent the phenomena that produce displacements and deformations in the continuum. As a result, the meaning of any equation formed
from these quantities can be interpreted visually by comparing it to the
concepts of continuum mechanics. This transparency provides insights
into the finite element method and its capabilities.

The insights provided by this transparency are important in two ways.
First, all aspects of the finite element method are simplified and extended.
These improvements occur because the theory of continuum mechanics
and the solution techniques are directly connected by the notation. Second,
these insights provide new research opportunities for both new and experienced users of the finite element method.

Research opportunities for readers new to the finite element method
exist because the use of the physically interpretable notation does much
to level the playing field between the novice and the expert. In contrast to
other advances, this improvement simplifies the finite element method and
provides the capacity for developments that were not previously available.
Thus, these possibilities are equally accessible to both new and experienced researchers.

3.2 SIGNIFICANCE OF THE NOTATIONAL CHANGE

The inclusion of physical meaning into the displacement interpolation polynomials provides the following specific results and research
opportunities.

- The strain modeling characteristics of individual elements can be evaluated by visual inspection. As we will see, the limitations and capabilities of individual elements can be identified during the element formulation process. This a priori evaluation means that extensive experience with finite element models is not needed to gain an understanding of the modeling capabilities of an element.
- The element stiffness matrix formulation procedure is improved and simplified. A convoluted approximate integration scheme is replaced by an exact integration technique. A significant part of this simplification consists of the elimination of a coordinate transformation that can introduce strain modeling errors into the finite element model. As a result of this development, the finite element method can be easily introduced to undergraduates.
- The computational effort required to form a stiffness matrix is reduced because fewer integrals must be evaluated. For example, the number of integrals that must be evaluated for a six-node element is reduced from 78 to 6.
- New research opportunities for improving error estimators are available because the errors can be estimated in terms of quantities that have direct significance in solid mechanics problems. The errors can be estimated in terms of pointwise values of stresses or strains. This contrasts with the commonly used metrics that are aggregated or averaged quantities. For example, the pointwise values allow estimates to be related to the failure criteria for a specific material.
- New research opportunities for improving mesh refinement guides are made possible. The number of elements needed to reduce the error to a predefined level can be identified by comparing an estimate of the exact solution with the modeling capabilities of the individual elements. This rational approach to mesh refinement replaces the correlations with error estimates that are widely used.

3.3 OBJECTIVES

This chapter has two primary objectives. The first is to put the physically interpretable notation on a solid theoretical foundation. This is accomplished by recognizing that the interpolation polynomials are truncated Taylor series expansions. Then, the Taylor series coefficients are then interpreted in terms of rigid body motion and strain quantities.

The second objective is to clarify the meaning of the individual coefficients and to demonstrate the usefulness of the notation. This is accomplished with two examples that use the notation to identify modeling characteristics that would have otherwise gone unrecognized. In the first example, the modeling capabilities of two finite elements are evaluated. In the second example, a counterintuitive characteristic of the hexagonal truss that was formed in the previous chapter is identified.

Both examples provide background for the element stiffness matrix formulation procedure to be presented in the next chapter. The first example introduces the strain representations used to form the element. The second example develops a coordinate transformation that introduces the nodal displacements into the analysis.

3.4 FINITE ELEMENT DISPLACEMENT INTERPOLATION POLYNOMIALS

Before proceeding to a discussion of the notation, we will demonstrate a procedure for identifying the algebraic terms needed for the displacement interpolation polynomial of a specific finite element. These polynomials are basic to individual finite elements in two ways. On the functional level, they implicitly define the modeling capabilities of an element. On the computational level, they are used to form a coordinate transformation that is required for computing the stiffness matrix for an element.

A modified form of Pascal's triangle is used to identify the algebraic terms needed for a specific nodal configuration. The standard form of Pascal's triangle shown in Figure 3.2a is modified by rotating it as shown in Figure 3.2b. This orientation aligns the x's and y's of the triangle with the horizontal and vertical x and y axes so that the nodal pattern of an element can be conveniently superimposed on the triangle.

In Figure 3.3, the nodal configurations for six- and nine-node elements are superimposed on the rotated form of Pascal's triangle. As can be

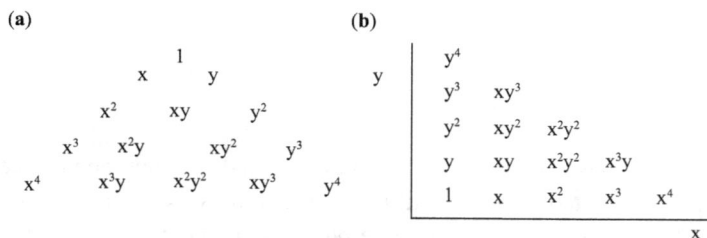

Figure 3.2. Two forms of Pascal's triangle: (a) standard form and (b) rotated form.

(a) **(b)**

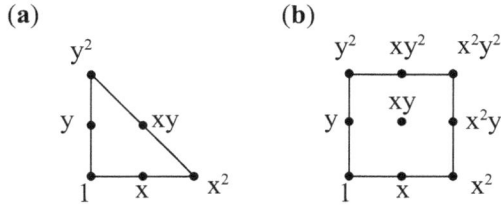

Figure 3.3. Two nodal configurations: (a) six-node triangle and (b) nine-node rectangle.

seen, each node is associated with a polynomial term that will be included in the displacement interpolation functions. Note that the nodal configurations are not required to have the spacing that is shown. That is to say, the superposition is topologically correct, not geometrically accurate.

For the case of the six-node triangle, the polynomials representing the displacements in the x and y directions are the following:

$$u(x,y) = a_1 + a_2x + a_3y + a_4x^2 + a_5xy + a_6y^2$$
$$v(x,y) = b_1 + b_2x + b_3y + b_4x^2 + b_5xy + b_6y^2$$

(3.1)

where u and v are the displacements in the x and y directions, respectively.

The two displacement representations contained in Equation 3.1 are complete second-order polynomials with arbitrary or content-free coefficients. Specifically, these coefficients have no direct meaning with respect to continuum mechanics. For example, we cannot recognize that the term a_2 represents the magnitude of the normal strain ε_x by visual inspection.

In the second example, we will identify the interpolation polynomials for the nine-node element shown in Figure 3.3b. When the terms associated with each node are included in the interpolation polynomial, we have the following:

$$u(x,y) = a_1 + a_2x + a_3y + a_4x^2 + a_5xy + a_6y^2 + a_7x^2y + a_8xy^2 + a_9x^2y^2$$
$$v(x,y) = b_1 + b_2x + b_3y + b_4x^2 + b_5xy + b_6y^2 + b_7x^2y + b_8xy^2 + b_9x^2y^2$$

(3.2)

Note that Equation 3.2 consists of augmented versions of the six-node interpolation polynomials given by Equation 3.1. Both the u and v components contain three additional terms, namely, x^2y, xy^2, and x^2y^2. In order to reinforce the fact that the arbitrary coefficients have no direct connection to solid mechanics, the reader might attempt to identify the strain quantities represented by the coefficients of these additional terms as an exercise.

3.5   PHYSICALLY INTERPRETABLE NOTATION

In this section, the physical meanings of the arbitrary coefficients contained in Equation 3.2 with respect to continuum mechanics are identified with a two-step process. In the first step, the interpolation polynomials are recognized as Taylor series expansions.

When the displacement interpolation polynomials for the nine-node configuration given by Equation 3.2 are written as truncated Taylor series expansions, we have the following:

$$u(x,y) = (u)_0 + (\partial u/\partial x)_0 x + (\partial u/\partial y)_0 y + (\partial^2 u/\partial x^2)_0 x^2 + (\partial^2 u/\partial x \partial y)_0 x y +$$
$$(\partial^2 u/\partial y^2)_0 y^2 + (\partial^3 u/\partial x^2 \partial y)_0 x^2 y + (\partial^3 u/\partial x \partial y^2)_0 x y^2 + (\partial^4 u/\partial x^2 \partial y^2)_0 x^2 y^2$$

$$(3.3a)$$

$$v(x,y) = (v)_0 + (\partial v/\partial x)_0 x + (\partial v/\partial y)_0 y + (\partial^2 v/\partial x^2)_0 x^2 + (\partial^2 v/\partial x \partial y)_0 x y +$$
$$(\partial^2 v/\partial y^2)_0 y^2 + (\partial^3 v/\partial x^2 \partial y)_0 x^2 y + (\partial^3 v/\partial x \partial y^2)_0 x y^2 + (\partial^4 v/\partial x^2 \partial y^2)_0 x^2 y^2$$

$$(3.3b)$$

The recognition of the polynomials as Taylor series expansions is significant because the arbitrary coefficients contained in Equation 3.2 have been replaced with terms having physical meaning. The coefficients are now expressed in terms of displacements and their derivatives. The subscript zeros on the coefficients mean that the coefficients are evaluated at the local origin of the Taylor series expansion.

In the second step, the Taylor series coefficients are transformed to rigid body motions and strain quantities. This is significant because the coefficients are now expressed in terms of variables that produce the displacements and deformations in the continuum.

When the Taylor series coefficients of Equation 3.3 are expressed in terms of rigid body and strain quantities, the result is the following (Boresi, Schmidt, and Sidebottom 1993; Dow 1999):

$$u(x,y) = (u_{rb})_0 + (\varepsilon_x)_0 x + (\gamma_{xy}/2 - r_{rb})_0 y + 1/2(\varepsilon_{x,x})_0 x^2 + (\varepsilon_{x,y})_0 x y$$
$$+ 1/2(\gamma_{xy,y} - \varepsilon_{y,x})_0 y^2 + 1/2(\varepsilon_{x,xy})_0 x^2 y + 1/2(\varepsilon_{x,yy})_0 x y^2 + 1/4(\varepsilon_{x,xyy})_0 x^2 y^2$$

$$v(x,y) = (v_{rb})_0 + (\gamma_{xy}/2 + r_{rb})_0 x + (\varepsilon_y)_0 y + 1/2(\gamma_{xy,x} - \varepsilon_{x,y})_0 x^2$$
$$+ (\varepsilon_{y,x})_0 x y + 1/2(\varepsilon_{y,y})_0 y^2 + 1/2(\varepsilon_{y,xx})_0 x^2 y + 1/2(\varepsilon_{y,xy})_0 x y^2 + 1/4(\varepsilon_{y,xxy})_0 x^2 y^2$$

$$(3.4)$$

As can be seen, the arbitrary coefficients of Equation 3.2 have been replaced by coefficients that have specific physical meaning in solid mechanics. The zeroth order coefficients are constants that represent the rigid body displacements $(u_{rb})_0$ and $(v_{rb})_0$. The four first-order coefficients of the x and y terms $(\varepsilon_x)_0$, $(\varepsilon_y)_0$, and, $(\gamma_{xy})_0$ represent constant strains and the rigid body rotation $(r_{rb})_0$ (see note 1 at the end of the chapter).

The higher-order terms represent gradients of the strain components. For example, the coefficients of the xy terms represent the rate of change of the normal strains in the coordinate directions. The term $(\varepsilon_{x,y})_0$ represents the change in $(\varepsilon_x)_0$ in the y direction. Analogously, the term $(\varepsilon_{y,x})_0$ represents the change in $(\varepsilon_y)_0$ in the x direction.

3.6 VALIDATION OF THE PHYSICALLY INTERPRETABLE COEFFICIENTS

In this section, we will demonstrate that Equations 3.3 and 3.4 are equivalent with four examples. As noted earlier, all of the physically interpretable coefficients are derived in detail in Dow (1999, 2012).

The meaning of the leading constant term of Equation 3.3a, $(u)_0$, is clear. It represents the displacement of the origin of the Taylor series expansion. In order to directly relate this term to continuum mechanics, we will interpret it as the rigid body displacement in the x direction and express it as $(u_{rb})_0$. In this interpretation, each point in the element is considered to move the same distance in the x direction as the origin.

The coefficient of the x term of Equation 3.3a, $(\partial u/\partial x)_0$, is a gradient term that represents the rate of change of the displacement in the x direction. In the context of solid mechanics, the normal strain in the x direction, $(\varepsilon_x)_0$, is defined by this term. When the coefficient of the x term is replaced by $(\varepsilon_x)_0$, this expression now relates directly to continuum mechanics.

Another coefficient that is relatively easy to interpret in terms of strain is the coefficient of x^2 in Equation 3.3a, namely, $(\partial^2 u/\partial x^2)_0$. This quantity can be interpreted as the first derivative of ε_x with respect to x or $(\partial \varepsilon_x/\partial x)_0$. In other words, this term express the rate of change in ε_x in the x direction at the local origin. In order to compress the notation, this gradient will be written more compactly as $(\varepsilon_{x,x})_0$. In this case, the subscript x after the comma indicates a derivative with respect to x.

A term whose physical meaning might not be obvious at first glance is the coefficient of the y term in Equation 3.3a, namely, $(\partial u/\partial y)_0$. This quantity can be related directly to solid mechanics by introducing the definitions of rigid body rotation around the z axis and shear strain in the x-y plane from linear elasticity. These quantities are defined as follows:

$$r_{rb} = \frac{1}{2}\left(\frac{\partial v}{\partial x} - \frac{\partial u}{\partial y}\right)$$

$$\gamma_{xy} = \frac{\partial u}{\partial y} + \frac{\partial v}{\partial x}$$

(3.5)

As can be seen, the rigid body rotation, r_{rb}, and the shear strain, γ_{xy}, are formed from the same two partial derivatives, $\partial u/\partial y$ and $\partial v/\partial x$. Both derivatives are present in the Taylor series expansion. The derivative $\partial u/\partial y$ is the coefficient of the y term in Equation 3.3a and the derivative $\partial v/\partial x$ is the coefficient of the x term in Equation 3.3b.

The Taylor series coefficient of the y term in Equation 3.3a can be related directly to continuum mechanics with a linear combination of the definitions of rigid body rotation and shear strain that are given by Equation 3.5. The coefficient $\partial u/\partial y$ is formed by combining the rigid body rotation and the shear strain expressions as follows:

$$\gamma_{xy}/2 - r_{rb} = \frac{1}{2}\left(\frac{\partial u}{\partial y} + \frac{\partial v}{\partial x}\right) - \frac{1}{2}\left(\frac{\partial v}{\partial x} - \frac{\partial u}{\partial y}\right)$$

$$= \frac{1}{2}\left(\frac{\partial u}{\partial y} + \frac{\partial v}{\partial x} - \frac{\partial v}{\partial x} + \frac{\partial u}{\partial y}\right)$$

$$= \frac{1}{2}\left(2\frac{\partial u}{\partial y}\right)$$

$$= \frac{\partial u}{\partial y}$$

(3.6)

Thus, the coefficient of the y term in Equation 3.3a becomes $(\gamma_{xy}/2 - r_{rb})_0$. In an analogous development, the coefficient of the x term in Equation 3.3b becomes $(\gamma_{xy}/2 + r_{rb})_0$.

3.7 EVALUATION OF A FOUR-NODE FINITE ELEMENT

In this demonstration of the capability of strain gradient notation, we will use the notation to evaluate the strain modeling characteristics of a four-node finite element. The strain representations are derived during the formulation of the stiffness matrix.

This early evaluation is significant because it means that the strain modeling characteristics do not have to be deduced from the solution of problems. In other words, the knowledge that surfaces during the element formulation process is available to beginning analysts as well as to experienced users of the finite element method.

The first step in this process is to form the strain representations for an individual finite element. The strain models are created by applying the definitions of strains to the displacement interpolation polynomials that have been identified for this element. Since the strain representations are then expressed in terms of strain quantities, we can evaluate their modeling characteristics by comparing the strain models to the theory of continuum mechanics.

The four-node element is chosen for evaluation because it contains several strain modeling errors. As we will see, the six-node element is chosen because it does not contain any modeling errors. Its only deficiency is due to the fact that it is limited to representing linear strains exactly. In other words, it cannot represent quadratic variations in the strains in the problem that it is attempting to represent. This limitation is inherent in the truncated interpolation polynomials used to form the linear strain element.

When the configuration of a four-node element is superimposed on a rotated Pascal's triangle, the displacement interpolation polynomials contain the following algebraic terms: 1, x, y, and xy. When we extract the components of Equation 3.4 that contain these terms, we have the following displacement interpolation polynomials:

$$u(x,y) = (u_{rb})_0 + (\varepsilon_x)_0 x + (\gamma_{xy}/2 - r_{rb})_0 y + (\varepsilon_{x,y})_0 x y$$
$$v(x,y) = (v_{rb})_0 + (\gamma_{xy}/2 + r_{rb})_0 x + (\varepsilon_y)_0 y + (\varepsilon_{y,x})_0 xy$$

$$(3.7)$$

The strain representations for a four-node element are found by applying the definitions of the three strain components from linear elasticity to Equation 3.7 to give the following:

$$\varepsilon_x(x,y) = (\partial u/\partial x) = (\varepsilon_x)_0 + (\varepsilon_{x,y})_0 y$$
$$\varepsilon_y(x,y) = (\partial v/\partial y) = (\varepsilon_y)_0 + (\varepsilon_{y,x})_0 x \qquad (3.8)$$
$$\gamma_{xy}(x,y) = (\partial v/\partial x + \partial u/\partial y) = (\gamma_{xy})_0 + (\varepsilon_{x,y})_0 x + (\varepsilon_{y,x})_0 y$$

Note that the two normal strain representations are truncated Taylor series expansions. However, they are not complete linear polynomials. The normal strain in the x direction does not contain an x term with the required coefficient $(\varepsilon_{x,x})_0$. In a similar manner, the representation of ε_y does not contain a y term with the required coefficient $(\varepsilon_{y,y})_0$.

In contrast, the shear strain representation contains both an x term and a y term. In spite of the fact the shear strain model is a complete polynomial, the four-node element representation of the shear strain representation contains strain modeling errors. Although, this element is capable of accurately representing constant strain because of the presence of the $(\gamma_{xy})_0$ term, this element, it cannot represent the linear variations in γ_{xy}. Specifically, the coefficients of the x and y terms should be $(\gamma_{xy,x})_0$ and $(\gamma_{xy,y})_0$, respectively.

The presence of the normal strain coefficients in the shear strain expression produces errors in the shear strain representation. In the terminology of the finite element method, the existence of the normal strain terms in the shear strain expression is called *shear locking*. Shear locking indicates that excess shear strain occurs when there is a contribution by the normal strain gradient terms, $(\varepsilon_{x,y})_0$ and $(\varepsilon_{y,x})_0$.

Next, we will further analyze the three strain representations by comparing them to the requirements prescribed by the constitutive relationship for plane stress. The constitutive relationship is the following:

$$\begin{Bmatrix} \sigma_x \\ \sigma_y \\ \tau_{xy} \end{Bmatrix} = \frac{E}{1-\upsilon^2} \begin{bmatrix} 1 & \upsilon & 0 \\ \upsilon & 1 & 0 \\ 0 & 0 & (1-\upsilon)/2 \end{bmatrix} \begin{Bmatrix} \varepsilon_x \\ \varepsilon_y \\ \gamma_{xy} \end{Bmatrix} \qquad (3.9)$$

The most obvious strain modeling error occurs in the shear strain representation shown in Equation 3.8. As mentioned earlier, this error of commission consists of the presence of the two normal strain terms in the expression. The existence of these terms constitutes a strain modeling error since the constitutive relationship for the shear given in Equation 3.9 does not contain any normal strain contributions.

The strain modeling errors in the normal strain representations are not as obvious as the errors in the shear strain representation. This is the case because these errors are errors of omission, not errors of commission. Since the normal strain in the x direction does not contain the Taylor series term $(\varepsilon_{x,x})_0$ x, it cannot represent the Poisson effect when the strain state $(\varepsilon_{y,x})_0$ x exists in the problem being modeled. This defect makes the four-node element overly stiff. An identical error exists in the representation of ε_y.

Furthermore, it is shown in Dow (1999) that the absence of these two terms in the normal strain representations causes the normal strain representation to vary with the orientation of the element. In other words, the normal strain representations are not invariant with the orientation of the element as should be the case.

In summary, we can conclude that a four-node element can only accurately represent constant strains on its domain. In other words, a four-node element cannot represent the strain distributions in the continuum any better than a three-node constant strain triangle.

In Dow (1999), it is shown that any four-sided element contains errors that are analogous to those discussed for the four-node element. The eight- and nine-node elements can represent linear strain distributions, but they contain errors in the higher-order strain representations that are similar to those identified for the four-node element.

3.8 EVALUATION OF A SIX-NODE FINITE ELEMENT

The strain modeling characteristics of the six-node element will now be evaluated. The displacement interpolation functions for the six-node triangular element shown in Figure 3.3a are embedded in the representation for the nine-node representation given by Equation 3.4. As can be seen in Equation 3.1, the displacement interpolation polynomial for a six-node element consists of complete quadratic representations.

When the complete quadratic terms are extracted from Equation 3.4, we have the displacement interpolation polynomials for a six-node finite element expressed in strain gradient notation:

$$u(x,y) = (u_{rb})_0 + (\varepsilon_x)_0 x + (\gamma_{xy}/2 - r_{rb})_0 y + 1/2(\varepsilon_{x,x})_0 x^2 + (\varepsilon_{x,y})_0 xy$$
$$+ 1/2(\gamma_{xy,y} - \varepsilon_{y,x})_0 y^2$$

$$v(x,y) = (v_{rb})_0 + (\gamma_{xy}/2 + r_{rb})_0 x + (\varepsilon_y)_0 y + 1/2(\gamma_{xy,x} - \varepsilon_{x,y})_0 x^2 + (\varepsilon_{y,x})_0 xy$$
$$+ 1/2(\varepsilon_{y,y})_0 y^2$$

$$(3.10)$$

The strain representations for a six-node element are found by applying the definitions of the three strain components from linear elasticity to Equation 3.10 to give the following:

$$\varepsilon_x(x,y) = \partial u/\partial x = (\varepsilon_x)_0 + (\varepsilon_{x,x})_0 x + (\varepsilon_{x,y})_0 y$$
$$\varepsilon_y(x,y) = \partial v/\partial y = (\varepsilon_y)_0 + (\varepsilon_{y,x})_0 x + (\varepsilon_{y,y})_0 y \qquad (3.11)$$
$$\gamma_{xy}(x,y) = (\partial v/\partial x + \partial u/\partial y) = (\gamma_{xy})_0 + (\gamma_{xy,x})_0 x + (\gamma_{xy,y})_0 y$$

As can be seen, the three strain components are represented by complete linear representations. In this case, the coefficients are those expected for a complete linear Taylor series expansion. The presence of the correct constant and linear terms in both of the normal strain components means that the Poisson effect is represented correctly.

The shear strain representation is a complete linear polynomial. In the six-node element, these are the correct shear strain terms. The linear terms are not normal strain terms as was the case for the four-node element.

We can conclude that a six-node element can accurately represent is a linear strain distribution. As might be expected, the six-node element is known as the linear strain triangle. If the actual strain distribution is more complex, the six-node element representation can only approximate the actual strain distribution.

The differences between the finite element representation and the exact solution are called discretization errors. This name is appropriate because the errors that exist are due to the attempt to represent an infinite degree-of-freedom problem with a discrete or finite number of degrees-of freedom.

3.9 EQUIVALENT CONTINUUM PARAMETERS

The purpose of this example is to give explicit meaning to the physically interpretable coefficients. In this example, interpolation polynomials expressed in strain gradient notation are used to extract quantities that

Twelve bar equilateral triangle hexogon

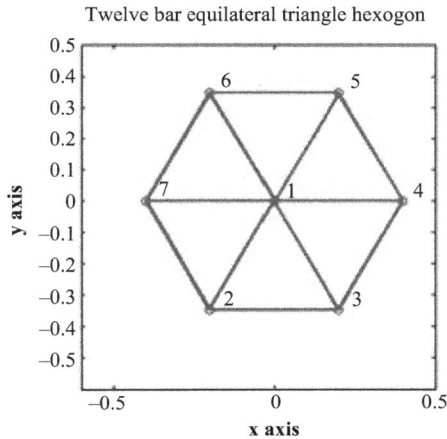

Figure 3.4. A repeated element.

represent Young's modulus, Poisson's ratio, and the shear modulus from the hexagonal truss that was developed in Chapter 2. The configuration of this truss which was originally shown in Figure 2.1 is repeated for convenience as Figure 3.4

The extraction of these *equivalent continuum parameters* identifies two counterintuitive characteristics of this truss. On the one hand, we will see that the equivalent continuum parameters are independent of its orientation. On the other hand, we will see that these equivalent continuum parameters are analogous to the constitutive relations for a plane stress problem. Later, we will see that these characteristics can be useful in the design of trusses.

The equivalent continuum parameters are computed by forming an analogous relationship to the strain energy expression for a continuous domain. The strain energy expression for a two-dimensional continuum is the following (Boresi 1965; Boresi, Schmidt, and Sidebottom 1993):

$$SE = \frac{1}{2} \int_\Omega \begin{Bmatrix} \varepsilon_x \\ \varepsilon_y \\ \gamma_{xy} \end{Bmatrix}^T \begin{bmatrix} C_{11} & C_{12} & C_{13} \\ C_{12} & C_{22} & C_{23} \\ C_{13} & C_{23} & C_{33} \end{bmatrix} \begin{Bmatrix} \varepsilon_x \\ \varepsilon_y \\ \gamma_{xy} \end{Bmatrix} d\Omega \qquad (3.12)$$

where the matrix containing the Cs is the constitutive relationship and Ω denotes the domain of the problem.

The discrete analog of Equation 3.12 is formed from the following expression for the strain energy contained in a discrete structure:

$$SE = 1/2 \ \{d\}^T [K] \ \{d\} \qquad (3.13)$$

where [K] is the unrestrained stiffness matrix for the discrete structure and {d} is the vector of nodal displacements.

In order to form a strain energy expression for the truss that is analogous to Equation 3.12, we must transform the nodal displacements in Equation 3.13 to strain quantities. This transformation is formed using the following displacement functions:

$$u(x,y)_i = (u_{rb})_0 + (\varepsilon_x)_0 x_i + (\gamma_{xy}/2 - r_{rb})_0 y_i$$
$$v(x,y)_i = (v_{rb})_0 + (\gamma_{xy}/2 + r_{rb})_0 x_i + (\varepsilon_y)_0 y_i$$

(3.14)

When Equation 3.14 is expressed in matrix form in preparation for forming the required coordinate transformation, we have the following:

$$\begin{Bmatrix} u(x,y)_i \\ v(x,y)_i \end{Bmatrix} = \begin{bmatrix} 1 & 0 & -y_i & x_i & 0 & y_i/2 \\ 0 & 1 & x_i & 0 & y_i & x_i/2 \end{bmatrix} \{\varepsilon,\}_0$$

(3.15)

where $\{\varepsilon,\}_0 = [u_{rb} \quad v_{rb} \quad r_{rb} \quad \varepsilon_x \quad \varepsilon_y \quad \gamma_{xy}]_0^T$

The six independent variables in equations 3.14 and 3.15 consist of the three rigid body motions, $(u_{rb})_0$, $(v_{rb})_0$, and $(r_{rb})_0$, and the three strain components, $(\varepsilon_x)_0$, $(\varepsilon_y)_0$, and $(\gamma_{xy})_0$.

The three rigid body motions are included in this demonstration for two reasons. On one hand, the rigid body motions are easy to visualize and, by definition, they produce no strain energy in a structure. On the other hand, as we shall see in the next chapter, the inclusion of the rigid body displacements simplifies the computation of the stiffness matrix of an element. Therefore, it is instructive to demonstrate how these quantities are exhibited in a discrete structure (see note 2).

The three strain gradient terms, $(\varepsilon_x)_0$, $(\varepsilon_y)_0$, and $(\gamma_{xy})_0$, are analogous to the pointwise strains, ε_x, ε_y, and γ_{xy}, that serve as the independent variables in Equation 3.12. Consequently, these quantities must be included in the coordinate transformation that will be used to change the independent variables in Equation 3.13 from nodal displacements to strain gradient quantities. This transformation allows the equivalent continuum parameters to be identified.

The required transformation is formed by substituting the coordinates of each node of the truss into the displacement interpolation functions given by Equation 3.15. Since the truss has seven nodes, the nodal coordinates for each node is substituted into Equation 3.15 to form the required coordinate transformation. When these substitutions are made and the equations are combined, the coordinate transformation has the following form:

$$
\begin{Bmatrix} u_1 \\ u_2 \\ u_3 \\ u_4 \\ u_5 \\ u_6 \\ u_7 \\ v_1 \\ v_2 \\ v_3 \\ v_4 \\ v_5 \\ v_6 \\ v_7 \end{Bmatrix}
=
\begin{bmatrix}
1 & 0 & -y_1 & x_1 & 0 & y_1/2 \\
1 & 0 & -y_2 & x_2 & 0 & y_2/2 \\
1 & 0 & -y_3 & x_3 & 0 & y_3/2 \\
1 & 0 & -y_4 & x_4 & 0 & y_4/2 \\
1 & 0 & -y_5 & x_5 & 0 & y_5/2 \\
1 & 0 & -y_6 & x_6 & 0 & y_6/2 \\
1 & 0 & -y_7 & x_7 & 0 & y_7/2 \\
0 & 1 & x_1 & 0 & y_1 & x_1/2 \\
0 & 1 & x_2 & 0 & y_2 & x_2/2 \\
0 & 1 & x_3 & 0 & y_3 & x_3/2 \\
0 & 1 & x_4 & 0 & y_4 & x_4/2 \\
0 & 1 & x_5 & 0 & y_5 & x_5/2 \\
0 & 1 & x_6 & 0 & y_6 & x_6/2 \\
0 & 1 & x_7 & 0 & y_7 & x_7/2
\end{bmatrix}
\begin{Bmatrix} u_{rb} \\ v_{rb} \\ r_{rb} \\ \varepsilon_x \\ \varepsilon_y \\ \gamma_{xy} \end{Bmatrix}_0
\tag{3.16}
$$

As can be seen in Equation 3.16, the transformation consists of six columns. Each column identifies the nodal displacements or shape that each of the strain gradient coefficients imposes on the truss. The value of the components of the strain gradient vector specifies the magnitude of the contribution that each of the shapes makes to the overall displacement of the structure. The nodal locations for the truss corresponding to y_i and x_i in Equation 3.16 are identified in Table 3.1.

For example, the second column of Equation 3.16 identifies the shape of the rigid body displacement in the y direction. The actual magnitude of the displacements is controlled by the size of the associated strain gradient coefficient. If the rigid body displacement coefficient in the y direction, $(v_{rb})_0$, is equal to, say, 0.10 units, the product of column 2 with this scalar

Table 3.1. Nodal coordinate locations

Node	x	y
1	0	0
2	−0.2	−0.3464
3	0.2	−0.3464
4	0.4	0
5	0.2	0.3464
6	−0.2	0.3464
7	−0.4	0

value indicates that each of the seven nodes will be displaced by 0.10 units in the y direction.

We will now illustrate the meaning of Equation 3.16 in Figure 3.5 by superimposing the nodal displacements associated with four of the strain gradient configurations on the original nodal locations of the truss. The rigid body displacement in the y direction discussed previously is shown superimposed on the original location of the truss in Figure 3.5a. As can be seen, each node of the truss moves 0.10 units in the positive y direction. The truss moves as one piece without deformation. This is the very definition of a rigid body motion.

The relative nodal displacements for a rigid body rotation in the x-y plane are contained in column 3 of Equation 3.16. If the truss is given a rigid body rotation, $(r_{rb})_0$, that is equal to a value of 0.20 units, the truss rotates in a counterclockwise direction without deformation with respect to its original position as shown in Figure 3.5b.

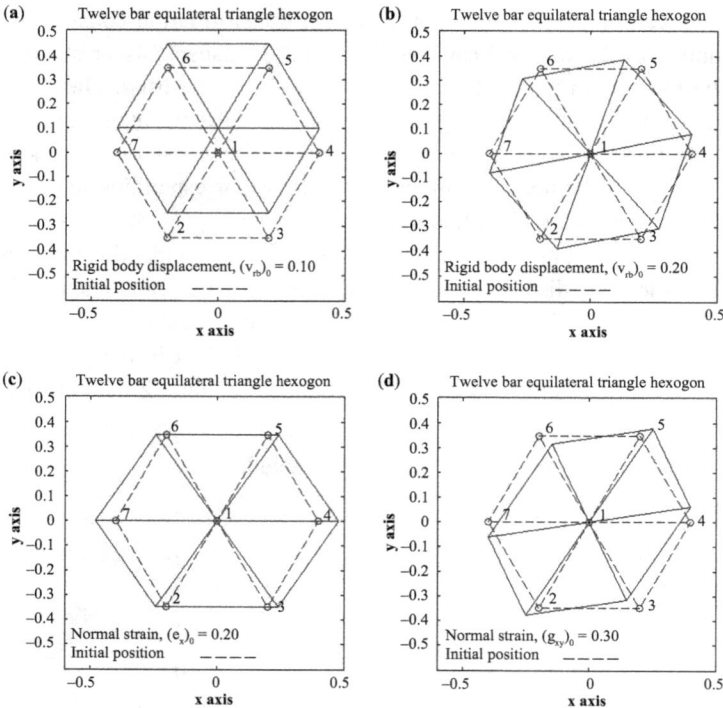

Figure 3.5. Strain state displacements and deformations: (a) a rigid body displacement, $(v_{rb})_0$; (b) a rigid body rotation, $(r_{rb})_0$; (c) a normal strain, $(\varepsilon_x)_0$; and (d) a shear strain, $(\gamma_{xy})_0$.

In contrast to the rigid body motions, the truss is deformed when a normal strain in the x direction, $(\varepsilon_x)_0$, is imposed on it. The displacement pattern for this strain condition is contained in column 4 of Equation 3.16. When $(\varepsilon_x)_0$ is positive, nodes move in such a way as to produce a deformation pattern that is equivalent to tension in the continuum. Conversely, if $(\varepsilon_x)_0$ is negative, the truss is compressed.

The displacements of the nodes relative to the original configuration are shown in Figure 3.5c when the truss is given a strain that is equal to 0.20 of a unit. As would be expected for a normal strain in the x direction, no movement of the nodes occurs in the y direction. Note that the local origin which is positioned at node 1 does not move. This is consistent with the fact that the local x value at the origin is equal to zero. As can be seen in the figure, the truss stretches along the x axis, so it is in tension.

As a final example, let us consider the shear strain. The displacement pattern for this deformation is contained in column 6 of Equation 3.16. When the truss is given a shear strain $(\gamma_{xy})_0$ that is equal to 0.30 units, the truss is deformed relative to its initial position as shown in Figure 3.5d. Note that the truss lengthens along one diagonal and shortens along the other diagonal. Such a pattern is characteristic of a shear deformation (see note 2).

Now that the meanings of the individual strain gradient coefficients have been demonstrated, we are in the position to extract the equivalent continuum parameters from the stiffness matrix of the truss. In preparation for computing the equivalent continuum parameters for the truss, let us put Equation 3.16 in the following compact form:

$$\{d\} = [T]\{\varepsilon,\}_0 \qquad (3.17)$$

where $\{d\}$ indicates the nodal displacements, $[T]$ identifies the transformation matrix, and $\{\varepsilon,\}_0$ designates the vector of strain gradient terms.

When the expression for the strain energy in discrete coordinates given by Equation 3.13 is transformed to strain gradient coordinates by substituting Equation 3.17, we have the following:

$$SE = 1/2 \{d\}^T[K] \{d\}$$

$$= 1/2 \{\varepsilon,\}_0^T [T]^T[K] [T]\{\varepsilon,\}_0 \qquad (3.18)$$

The equivalent continuum parameters of the truss are contained in the kernel of Equation 3.18 which is given by $[T]^T[K] [T]$. When the computation of the kernel identified in Equation 3.18 is performed for the hexagonal truss for the orientation shown in Figure 3.1 and the nodal coordinates given in Table 3.1, the result is the following:

$$[T]^T[K][T] = \begin{bmatrix} 0.0 & 0.0 & 0.0 & 0.0 & 0.0 & 0.0 \\ 0.0 & 0.0 & 0.0 & 0.0 & 0.0 & 0.0 \\ 0.0 & 0.0 & 0.0 & 0.0 & 0.0 & 0.0 \\ 0.0 & 0.0 & 0.0 & 18.0 & 6.0 & 0.0 \\ 0.0 & 0.0 & 0.0 & 6.0 & 18.0 & \sim 0.0 \\ 0.0 & 0.0 & 0.0 & 0.0 & \sim 0.0 & 6.0 \end{bmatrix} \tag{3.19}$$

As can be seen, the first three rows and columns which are associated with the rigid body motions are all equal to zero. This result is expected because a rigid body motion, by definition, does not deform the entity undergoing the motion. Therefore, no strain energy is produced. Even though this result is expected, it is significant. In the next chapter, we shall see that the presence of these rows and columns of zeros are instrumental in simplifying the formulation of the stiffness matrices for finite elements as well as reducing the number of integrals that must be evaluated.

The nonzero quadrant of Equation 3.19 is analogous to the strain energy of the two-dimensional continuum given by Equation 3.12. When we form the analogous strain energy expression for the truss, we have the following (Dow 2012):

$$SE = 1/2 \begin{Bmatrix} (\varepsilon_x)_0 \\ (\varepsilon_y)_0 \\ (\gamma_{xy})_0 \end{Bmatrix}^T \begin{bmatrix} C_{11} & C_{12} & C_{13} \\ C_{12} & C_{22} & C_{23} \\ C_{13} & C_{23} & C_{33} \end{bmatrix}_{Equiv} \begin{Bmatrix} (\varepsilon_x)_0 \\ (\varepsilon_y)_0 \\ (\gamma_{xy})_0 \end{Bmatrix} \tag{3.20}$$

In order to explicate the meaning of the equivalent continuum parameters, let us interpret the nonzero values in Equation 3.19 in terms of the plane stress problem. The constitutive relationship for a plane stress problem has the following form:

$$\begin{Bmatrix} (\sigma_x) \\ (\sigma_y) \\ (\tau_{xy}) \end{Bmatrix} = \frac{E}{(1-v^2)} \begin{bmatrix} 1 & v & 0 \\ v & 1 & 0 \\ 0 & 0 & \frac{1-v}{2} \end{bmatrix} \begin{Bmatrix} (\varepsilon_x) \\ (\varepsilon_y) \\ (\gamma_{xy}) \end{Bmatrix} \tag{3.21}$$

where E is Young's modulus, v is Poisson's ratio, and element (3, 3) is the shear modulus.

In Equation 3.19, the two elements that are indicated as being near zero are equal to 0.0002, which is very small when compared to the nonzero terms. Thus, we can treat this hexagonal truss as a solid medium with properties similar to, for example, steel or aluminum. This, in turn, means that we can treat a structure formed from repeated hexagonal truss elements as a plane stress problem.

When elements (1, 1) and (1, 2) of Equation 3.21 are equated to the equivalent elements of Equation 3.19, namely, 18.0 and 6.0, Young's modulus is found to be equal to 16.0 and Poisson's ratio is found to be equal to 1/3. As a further confirmation, when these values are inserted into element (3, 3) of Equation 3.21, the result is found to be 6.0. This is identical to the value for the corresponding element in Equation 3.19.

The fact that the constitutive relationship for the hexagonal truss is analogous to the constitutive relationship of an isotropic material implies that the truss possesses the same properties regardless of its orientation. At first, this concept seems implausible because of the different *shapes* that the truss presents to the x and y axes when it is rotated. In the next section, we will demonstrate that this truss has isotropic properties.

3.10 DEMONSTRATION OF ISOTROPIC EQUIVALENT CONTINUUM PARAMETERS

This section has two objectives. The first is to demonstrate the isotropic properties of the hexagonal truss. As a result of this characteristic, this hexagonal truss is designated as *isogrid* because iso- is the Greek prefix that means *the same*. The second objective is to show how the characteristics of isogrid can be useful in the design.

The isotropic nature of isogrid is demonstrated by extracting the equivalent continuum parameters for the truss after it has been rotated through a series of angles as shown in Figure 3.6. This rotation is accomplished by finding the new locations of the nodal coordinates after the rotation with the following coordinate transformation:

$$\left\{ \begin{matrix} x \\ y \end{matrix} \right\}_{new} = \begin{bmatrix} \cos\theta & -\sin\theta \\ \sin\theta & \cos\theta \end{bmatrix} \left\{ \begin{matrix} x \\ y \end{matrix} \right\}_{old} \tag{3.22}$$

When this transformation is applied for rotations of 5, 15, and 22.5 degrees and the equivalent continuum parameters are computed, we have the result shown in Figure 3.6. As noted on the figures, the values of the equivalent continuum parameters are equal to those given in Equation 3.19. In other words, the equivalent continuum parameters are identical regardless of the orientation of the truss.

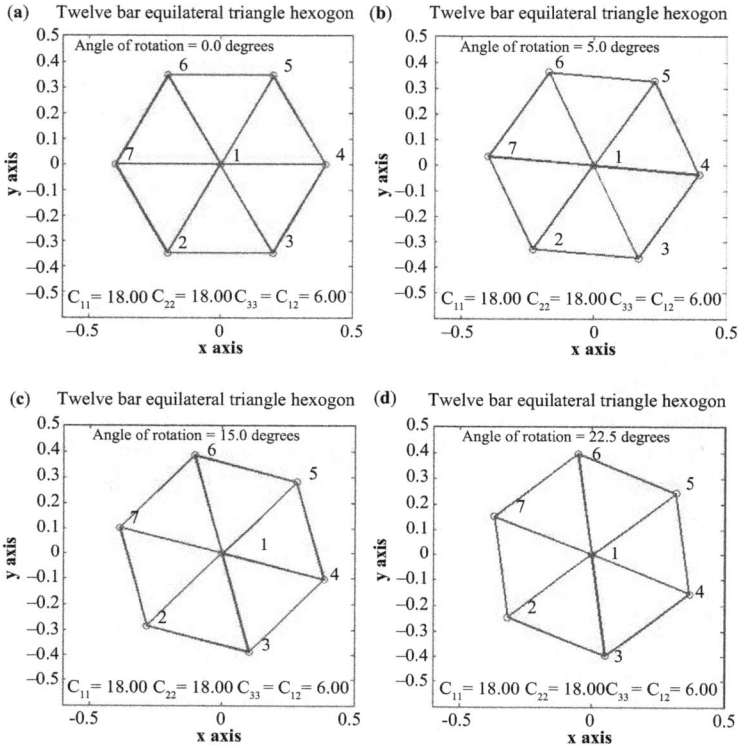

(a) Twelve bar equilateral triangle hexogon

(b) Twelve bar equilateral triangle hexogon

Angle of rotation = 0.0 degrees
$C_{11}= 18.00\ C_{22}= 18.00\ C_{33} = C_{12}= 6.00$

Angle of rotation = 5.0 degrees
$C_{11}= 18.00\ C_{22}= 18.00\ C_{33} = C_{12}= 6.00$

(c) Twelve bar equilateral triangle hexogon

(d) Twelve bar equilateral triangle hexogon

Angle of rotation = 15.0 degrees
$C_{11}= 18.00\ C_{22}= 18.00\ C_{33} = C_{12}= 6.00$

Angle of rotation = 22.5 degrees
$C_{11}= 18.00\ C_{22}= 18.00\ C_{33} = C_{12}= 6.00$

Figure 3.6. Hexagonal truss elements: (a) angle of rotation = 0 degrees, (b) angle of rotation = 5 degrees, (c) angle of rotation = 15 degrees, and (d) angle of rotation = 22.5 degrees.

In this example, we have seen that the hexagonal truss has the same structural characteristics in the x and y directions regardless of it orientation. This means that the orientation of the hexagonal truss need not be considered when designing a structure based on this configuration. Thus, great flexibility in the design of structures that use the hexagonal truss as the essential building block is provided.

An example of a structure that uses the hexagonal truss as its basic element is shown in Figure 3.7. One of the repeated hexagonal truss elements is highlighted in this figure. The invariance of the equivalent continuum parameters for the hexagonal truss provides another interesting capability. This invariance means that a structure formed from the repeated use of this hexagonal element, such as that shown in Figure 3.7, can be analyzed as if it were a continuum.

In other words, the equivalent continuum parameters can be used in the governing differential equations from continuum mechanics to

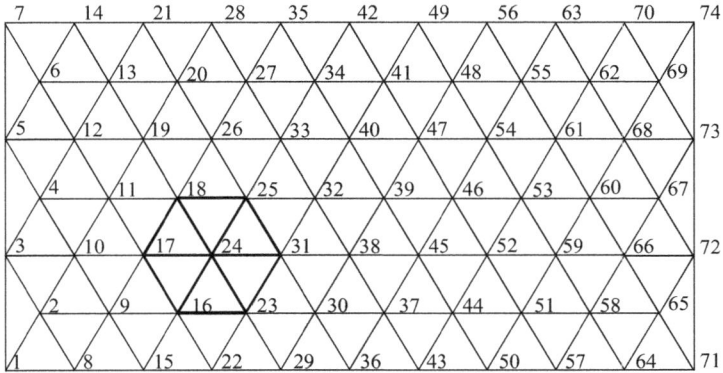

Figure 3.7. A truss formed from repeated hexagonal elements.

analyze the overall behavior of a structure formed from the repeated use of this truss element. Since a discrete structure is being analyzed as if it were a continuous problem, this procedure can be viewed as an inversion of the finite element method (see note 3).

The ability to analyze a truss that is constructed from repeated hexagonal truss elements means that the displacements of the discrete structure can be approximated with the governing differential equations for plane stress. The governing differential equations are the following:

$$\left(\frac{E}{1-v^2}\right)\left[u_{,xx} + vv_{,xy} + \left(\frac{1-v}{2}\right)\left(u_{,yy} + v_{,xy}\right)\right] = f_x$$

$$\left(\frac{E}{1-v^2}\right)\left[v_{,yy} + vu_{,xy} + \left(\frac{1-v}{2}\right)\left(v_{,xx} + u_{,xy}\right)\right] = f_y$$

(3.23)

where f_x and f_y are the applied loads, and the material properties are the equivalent continuum parameters.

In order to reaffirm the central role of strain quantities in solid mechanics, the governing differential equations for the plane stress problem given in Equation 3.23 can be expressed directly in terms of strain gradient quantities as follows:

$$\varepsilon_{x,x} + v\varepsilon_{y,x} + \left(\frac{1-v}{2}\right)(\gamma_{xy,y}) = -\left(\frac{1-v^2}{E}\right)f_x$$

$$\varepsilon_{y,y} + v\varepsilon_{x,y} + \left(\frac{1-v}{2}\right)(\gamma_{xy,x}) = -\left(\frac{1-v^2}{E}\right)f_y$$

(3.24)

As can be seen, these strain quantities are identical to terms in strain gradient notation.

3.11 SUMMARY

In this chapter, the difficult task of convincing readers to embrace a change in notation was approached both directly and indirectly. In the direct approach, the physically interpretable notation was given a solid theoretical foundation by showing it to be a specialized Taylor series expansion. Then the relationship of the notation to solid mechanics was demonstrated with two applications.

The first application demonstrated the direct connection to the finite element method by using the notation to evaluate the modeling capabilities of four- and six-node elements. The second application extracted equivalent continuum properties from a truss structure to show that the notation was closely related to the constitutive relations for solid mechanics.

The indirect approach identified the advantages of the new notation and outlined the developments to come in later chapters. The primary advantages outlined consist of (1) simplification of the element formulation procedure, (2) visual identification of the modeling capabilities of individual elements, and (3) new research opportunities provided by the notation.

While valuable to those having extensive experience with the finite element method, the primary advantage of the development presented here is to those new to computational mechanics. This is the case for three reasons. In the first place, the resulting simplification allows the finite element method to be presented at an earlier stage in the educational process (see note 4). Moreover, the physical insights into the modeling capabilities of individual elements reduce the need for extensive experience to understand the behavior of finite element models. Finally, since up to now the use of this notation has not been embraced as the standard approach, the ability to develop improved error estimators and refinement guides is available to anyone with this knowledge regardless of their level of experience.

3.12 NOTES

1. The ability of a planar element to represent the three rigid body motions and the three constant strain states, ε_x, ε_y, and γ_{xy}, means that a model can represent the continuum accurately as it is refined

regardless of the number of elements it takes. In the limit, this means that the element can represent the motion and strains at individual points. The ability to represent these conditions constitutes the *convergence criteria* for an individual element.

2. The classic example of a shear deformation is a sagging gate. When the gate sags, one diagonal gets longer and the other gets shorter. The constraint against shear in a gate is the diagonal element in a gate.

3. The extraction and use of equivalent continuum parameters and further references to the procedure are contained in Chapter 6 of Dow (1999).

4. As a test of the ability of this notation to make the finite element method available to undergraduates, I had a freshman Introduction to Engineering Computing class to develop finite element stiffness matrices using strain gradient notation. They then solved one-dimensional problems with MATLAB. Of course, they were not told that they were doing anything that might be considered difficult. As far as they knew, they were applying interpolation polynomials that matched the problem being solved.

3.13 EXERCISES

1. With a figure based on the rotated form of Pascal's triangle show that the displacement interpolation polynomials for a four-node rectangle contain the following terms: 1, x, y, and xy.

2. Show that $\varepsilon_{x,xx}$ is identical to the Taylor series coefficient d^3u/dx^3.

3. Reduce the displacement interpolation function given for u in Equation 3.6 or Equation 3.10 to one dimension, that is, retain only the x terms. Then, plot the nodal displacements for a three-node bar for a unit value of each of the strain states. Place the origin on the interior node with the positive x axis pointing to the right.

4. Redo exercise 3 with the origin located on the left hand node of the bar.

CHAPTER 4

AN IMPROVED STIFFNESS MATRIX FORMULATION PROCEDURE

4.1   INTRODUCTION

This chapter presents an improved procedure for forming finite element stiffness matrices. The improvements take advantage of the use of the physically interpretable, strain gradient notation that was developed in the last chapter. This improved element stiffness matrix formulation procedure *renders the widely used isoparametric element formulation procedure obsolete in two ways.*

First, the improved approach simplifies the formulation process and makes it easier to understand. Second, the improved approach does not contain the strain modeling errors that exist in some isoparametric elements (Dow 1999).

Both improvements are made possible because a step that is basic to the isoparametric procedure is eliminated. In this step, the actual element geometry is mapped onto a regular shape. This is done in order to facilitate the numerical integrations that exist in the isoparametric formulation process. Hughes (2000), Argyris and Kelsey (1960), Tiag (1961) provide background concerning the isoparametric element formulation procedure.

An example of this mapping is depicted in Figure 4.1. The general quadrilateral element with four, eight, or nine nodes is depicted in Figure 4.1a. This element is mapped onto the square shown as Figure 4.1b. As can be seen, the different size quadrilateral subdivisions of the element are mapped onto the equal size subsquares of Figure 4.1b.

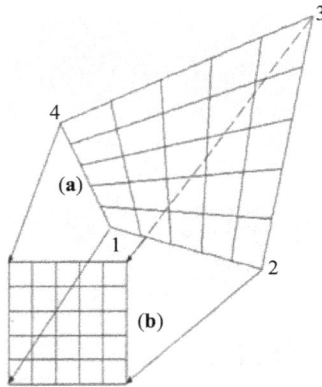

Figure 4.1. A 2-D isoparametric mapping.

However, this accommodation that facilitates the numerical integration is not without penalties. As a result of this mapping, the integrals that must be integrated are made more complicated. More significantly, strain modeling errors are introduced into isoparametric elements when the ratios of the subareas in the element and the square are not constant as is the case in Figure 4.1 (see note 1).

Examples of the strain modeling errors that exist in triangular six-node isoparametric elements that possess curved boundaries are shown in Figure 4.2 with contour plots. In Figure 4.2a, an element with a straight edge representing the strain state $\varepsilon_{x,y}$ does not contain any error in the representation of ε_x which varies linearly across the element in the y direction. This is the case because the triangles in the actual element map onto similar triangles that are used in the numerical integrations.

In contrast, the elements with curved boundaries shown in Figures 4.2b and 4.2c do not have regular shapes. The bottom edges are curved by moving the center node toward the center by a fraction of the length of the side, L. As a result of this deviation from a triangle, the isoparametric procedure introduces strain modeling errors into these elements.

The element shown in Figure 4.2b contains a significant maximum error of −26.9 percent in the representation of ε_x. When the curvature on the boundary is increased as shown in Figure 4.2c, the maximum error in ε_x expands to −77.3 percent.

In addition to improving the element formulation process, the use of the physically interpretable notation provides the basis for new classes of error estimators and refinement guides. Error estimators and refinement

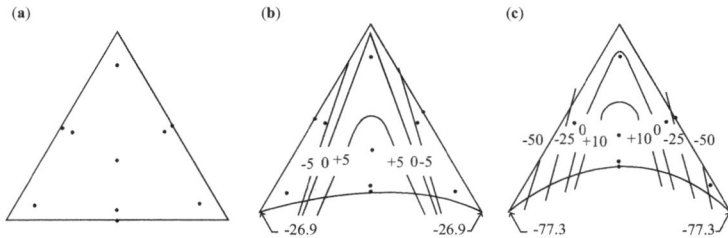

Figure 4.2. Six-node elements with increasing curvature: (a) $\Delta = 0.0$, (b) $\Delta = 0.1$ L, and (c) $\Delta = 0.2$ L.

guides are developed in Chapters 7 and 8, respectively. These capabilities comprise the heart of the adaptive refinement process.

4.2   OBJECTIVES

The primary objective of this chapter is to present the details for forming finite element stiffness matrices by using the physically interpretable strain gradient notation. The theoretical and computational advantages of this approach are highlighted in the introduction.

On the pedagogical level, the advantages of these improvements cannot be quantified. As discussed earlier, the stiffness matrix formulation procedure based on the use of the physically interpretable notation is less complicated than the isoparametric approach. In other words, the alternate approach is easier to learn (see notes 1 and 2).

An additional benefit derives from using the physically interpretable notation. Strain gradient notation provides the basis for showing that the finite element method and the finite difference methods share a common basis. This is significant because it allows the error analysis procedures developed in Chapter 7 and the refinement guides developed in Chapter 8 to be put on a solid theoretical foundation.

4.3   AN OVERVIEW OF THE STIFFNESS MATRIX FORMULATION PROCEDURE

The stiffness matrices and the applied load vectors for individual finite elements are formed by applying the principle of minimum potential energy. The stiffness matrix and the applied load vector are embedded in the strain energy and work functions of the potential energy expression, respectively. As a result, the stiffness matrix and the applied load vector can be formed in separate operations. The procedure for forming the stiffness matrix is presented here. Since the formulation of the applied load vector is so simple, it is not discussed here in detail.

The steps for forming the finite element stiffness matrix using strain gradient notation are presented in Table 4.1. The individual steps are as follows:

Step 1 identifies the terms of the interpolation polynomials and, hence, the strain gradient quantities that serve as the independent variables.

Step 2 forms and evaluates the strain representations that are derived from the displacement interpolation functions identified in Step 1.

Step 3 forms the strain energy expression in terms of strain gradient variables and identifies the small number of integrals that must be evaluated.

Step 4 evaluates the integrals identified in Step 3.

Step 5 forms the transformation from strain gradient coordinates to nodal coordinates in a two-step process. This transformation contains the finite difference derivative operators.

Step 6 substitutes the transformation created in step 5 into the strain energy expression formed in steps 3 and 4 to put the strain energy expression in terms of nodal coordinates.

Step 7 is a formality that is not actually performed in practice because the stiffness matrix is directly available by inspection in Step 6.

It should be noted that steps 1 and 2 are identical to operations performed in Chapter 3 to evaluate the modeling capabilities of the four- and

Table 4.1. Strain gradient based stiffness matrix formulation procedure

1. Identification of the interpolation polynomials.
2. Formulation and evaluation of the strain model.
3. Formulation of strain energy expression in strain gradient coordinates.
4. Integration of strain energy terms.
5. Formulation of strain gradient to nodal displacement transformation.
6. Transformation of strain energy expression to nodal coordinates.
7. Creation of the stiffness matrix by application of the principle of minimum potential energy to the strain energy expression.

six-node elements. In addition, steps 5 and 6 are similar to the operations performed in Chapter 3 to identify the equivalent continuum parameters for the hexagonal truss. In Chapter 3, the strain energy for the discrete structure is transformed from nodal displacements to strain gradient quantities in order to obtain an expression that is similar to the strain energy expression in the continuum.

4.4   AN EXAMPLE OF THE STRAIN GRADIENT FORMULATION PROCEDURE

The element stiffness matrix formulation procedure is demonstrated with the six-node element. This element is chosen because it is the simplest element with which to highlight the advantages of the strain gradient approach to element formulation over the isoparametric procedure.

This element requires the evaluation of only six integrals and it accurately represents the strains in an element with curved edges. The isoparametric version of this element requires the evaluation of 78 integrals and, as discussed in the introduction, introduces significant errors in the representations of the higher-order strain states if an element has a curved edge (see note 3).

STEP 1—POLYNOMIAL IDENTIFICATION

Three examples of six-node triangles with different configurations are shown in Figure 4.3. As can be seen, all three elements are represented by the same polynomial terms. These terms are identified by topologically superimposing the shape of these elements on the modified form of Pascal's triangle which was shown in Figure 3.1b.

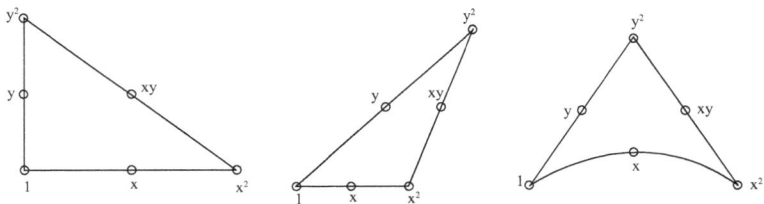

Figure 4.3. Three six-node triangles.

The displacement interpolation polynomials for the six-node element were identified in Equation 3.10. This equation is reproduced here for the convenience of the reader:

$$u(x,y) = (u_{rb})_0 + (\varepsilon_x)_0 x + (\gamma_{xy}/2 - r_{rb})_0 y + 1/2(\varepsilon_{x,x})_0 x^2$$
$$+ (\varepsilon_{x,y})_0 xy + 1/2(\gamma_{xy,y} - \varepsilon_{y,x})_0 y^2$$

$$(4.1)$$

$$v(x,y) = (v_{rb})_0 + (\gamma_{xy}/2 + r_{rb})_0 x + (\varepsilon_y)_0 y + 1/2(\gamma_{xy,x} - \varepsilon_{x,y})_0 x^2$$
$$+ (\varepsilon_{y,x})_0 xy + 1/2(\varepsilon_{y,y})_0 y^2$$

These complete second-order polynomials contain the following 12 linearly independent strain states that a six-node element can represent:

$$(u_{rb})_0 \ (v_{rb})_0 \ (r_{vb})_0 \ (\varepsilon_x)_0 \ (\varepsilon_y)_0 \ (\gamma_{xy})_0$$
$$(\varepsilon_{x,x})_0 \ (\varepsilon_{x,y})_0 \ (\varepsilon_{y,x})_0 \ (\varepsilon_{y,y})_0 \ (\gamma_{xy,x})_0 \ (\gamma_{xy,y})_0$$

$$(4.2)$$

The first six terms guarantee that the element can represent the rigid body motions and the constant strains states, which satisfy the convergence criteria for a finite element. This means that the element can represent a problem exactly if it is refined sufficiently. The second six terms represent the linear variations of the three strain components in the two coordinate directions.

STEP 2—STRAIN FORMULATION AND EVALUATION

The finite element strain representations are found by taking the appropriate derivatives of the displacement interpolation polynomials given in Equation 4.1. These strain models were formed in Chapter 3 where they were presented as Equation 3.11.

These representations are reproduced here for the convenience of the reader as follows:

$$\varepsilon_x(x,y) = \partial u/\partial x = (\varepsilon_x)_0 + (\varepsilon_{x,x})_0 x + (\varepsilon_{x,y})_0 y$$
$$\varepsilon_y(x,y) = \partial v/\partial y = (\varepsilon_y)_0 + (\varepsilon_{y,x})_0 x + (\varepsilon_{y,y})_0 y$$
$$\gamma_{xy}(x,y) = (\partial v/\partial x + \partial u/\partial y) = (\gamma_{xy})_0 + (\gamma_{xy,x})_0 x + (\gamma_{xy,y})_0 y$$

$$(4.3)$$

As discussed in Chapter 3, these representations contain no inherent modeling errors. However, the most complex strain distribution that these truncated Taylor series representations can model are linear strain conditions.

STEP 3—FORMULATION OF THE STRAIN ENERGY EXPRESSION

The strain energy expression for the continuous form of the plane stress problem is as follows:

$$SE = \frac{1}{2}\int_\Omega \{\varepsilon\}^T [E]\{\varepsilon\}\, d\Omega$$

where

$$\{\varepsilon\}^T = \begin{bmatrix} \varepsilon_x & \varepsilon_y & \gamma_{xy} \end{bmatrix}$$

$$[E] = \frac{E}{(1-v^2)}\begin{bmatrix} 1 & v & 0 \\ v & 1 & 0 \\ 0 & 0 & \frac{1-v}{2} \end{bmatrix} \tag{4.4}$$

$$E = \text{Young's modulus}$$

$$v = \text{Poisson's ratio}$$

The strain energy expression for the six-node element is formed by transforming the continuous strains contained in Equation 4.4 to the discrete finite element strain approximations contained in Equation 4.3. When the transformation from the continuous strains $\{\varepsilon\}$ to the discrete strain gradient quantities $\{\varepsilon,\}$ is formed, we have the following:

$$\{\varepsilon\} = [T]\{\varepsilon,\} \tag{4.5}$$

where

$$\{\varepsilon,\}^T = [(u_{rb})_0\ (u_{rb})_0\ (r_{rb})_0\ (\varepsilon_x)_0\ (\varepsilon_y)_0\ (\gamma_{xy})_0\ (\varepsilon_{x,x})_0\ (\varepsilon_{y,x})_0\ (\gamma_{xy,x})_0$$
$$(\varepsilon_{x,y})_0\ (\varepsilon_{y,y})_0\ (\gamma_{xy,y})_0\ ]$$

$$[T] = [[T_0][T_\varepsilon]]$$

$$[T_0] = \begin{bmatrix} 0 & 0 & 0 \\ 0 & 0 & 0 \\ 0 & 0 & 0 \end{bmatrix} \quad \text{and} \quad [T_\varepsilon] = \begin{bmatrix} 1 & 0 & 0 & x & 0 & 0 & y & 0 & 0 \\ 0 & 1 & 0 & 0 & x & 0 & 0 & y & 0 \\ 0 & 0 & 1 & 0 & 0 & x & 0 & 0 & y \end{bmatrix}$$

This strain representation differs slightly from the representation contained in Equation 4.3 because it is augmented with the *effects* of the rigid body motions. Although the rigid body motions do not add to the strain energy of the element, they are included in the transformation in order to facilitate the upcoming change from strain gradient coordinates to nodal displacements.

The contributions of the rigid body motions to the strain energy are appended to the beginning of the transformation by the partition designated as $[T_0]$. Since this matrix contains only zeros, the rigid body motions contribute nothing to the strains in the stiffness matrix being formed.

When the coordinate transformation given by Equation 4.5 is substituted into the strain energy expression given by Equation 4.4 and the independent variables are factored from the integral, we have the following:

$$
\begin{aligned}
SE &= \frac{1}{2}\int_\Omega \{\varepsilon,\}^T [T]^T [E][T]\{\varepsilon,\}\, d\Omega \\
&= \frac{1}{2}\{\varepsilon,\}^T \left[\int_\Omega [T]^T [E][T]\, d\Omega \right]\{\varepsilon,\} \\
&= \frac{1}{2}\{\varepsilon,\}^T \bar{U}\{\varepsilon,\}
\end{aligned}
\tag{4.6}
$$

It is in this step where the continuous problem is approximated with a discrete representation. We now have an approximation of the strain energy on the domain Ω which is represented by the finite element.

STEP 4—INTEGRATION OF THE STRAIN ENERGY TERMS

When the integrals that must be evaluated to form the stiffness matrix are extracted from Equation 4.6, we have the following for a region with a thickness of t:

$$
\begin{aligned}
\bar{U} &= \int_\Omega [T]^T [E]\, [T]\, d\Omega \\
&= \int_\Omega \begin{bmatrix} [T_0]^T [E][T_0] & [T_0]^T [E][T_\varepsilon] \\ [T_\varepsilon]^T [E][T_0] & [T_\varepsilon]^T [E][T_\varepsilon] \end{bmatrix} d\Omega
\end{aligned}
\tag{4.7}
$$

Since the matrix $[T_0]$ is the null matrix, any partition of Equation 4.7 containing this matrix is also a null matrix, that is, all of the terms are equal to zero. Consequently, we only have to integrate the nonzero terms contained in the fourth quadrant of Equation 4.7.

When the nonzero submatrix of Equation 4.7 is expanded, we have the following:

$$\bar{U}_{22} = \frac{t\,E}{(1-v^2)} \begin{bmatrix}
I_1 & v\,I_1 & 0 & I_2 & v\,I_2 & 0 & I_3 & I_3 & 0 \\
v\,I_1 & I_1 & 0 & v\,I & I_2 & 0 & v\,I & I_3 & 0 \\
0 & 0 & \alpha\,I_1 & 0 & 0 & \alpha\,I_2 & 0 & 0 & \alpha\,I_3 \\
I_2 & v\,I_2 & 0 & I_4 & v\,I_4 & 0 & I_5 & v\,I_5 & 0 \\
v\,I_2 & I_2 & 0 & v\,I_4 & I_4 & 0 & v\,I_5 & I_5 & 0 \\
0 & 0 & \alpha\,I_2 & 0 & 0 & \alpha\,I_4 & 0 & 0 & \alpha\,I_5 \\
I_3 & v\,I_3 & 0 & I_5 & v\,I_5 & 0 & I_6 & v\,I_6 & 0 \\
v\,I_3 & I_3 & 0 & v\,I_5 & I_5 & 0 & v\,I_6 & I_6 & 0 \\
0 & 0 & \alpha\,I_3 & 0 & 0 & \alpha\,I_5 & 0 & 0 & \alpha\,I_6
\end{bmatrix}$$

(4.8)

where

$$I_1 = \int_\Omega d\Omega \qquad I_2 = \int_\Omega x\,d\Omega \qquad I_3 = \int_\Omega y\,d\Omega$$
$$I_4 = \int_\Omega x^2\,d\Omega \qquad I_5 = \int_\Omega x\,y\,d\Omega \qquad I_6 = \int_\Omega y^2\,d\Omega$$

(4.9)

The six integrals contained in Equation 4.9 are the area, the two first moments of the area, and the three second moments of area. Since these integrals have a simple form, they can be integrated exactly. This contrasts to the 78 relatively complex integrals that must be evaluated in the isoparametric stiffness matrix formulation process with an approximation (see note 3).

Equations 4.7 and 4.8 contain significant characteristics that derive from the power and usefulness of the physically interpretable notation. As a consequence of explicitly including the rigid body terms in the displacement approximations, three quadrants of Equation 4.7 are known to contain only zeros from knowledge of continuum mechanics. As a result, these terms do not have to be integrated.

In the case of Equation 4.8, note that the integrals defined in Equation 4.9 appear more than once. The identification of these integrals in multiple locations in the strain energy expression means that the integrals only have to be evaluated once. This reduces the computational effort needed to form a stiffness matrix.

STEP 5—FORMULATION OF THE COORDINATE TRANSFORMATION MATRICES

The strain energy function given by Equation 4.8 is expressed in terms of the strain gradient quantities. However, the independent variables of the strain energy expressions must be expressed in terms of nodal displacements in order to assemble the individual stiffness matrices into a structure.

The transformation to nodal coordinates is produced in a two-step process. First, a transformation from nodal displacements to strain gradient variables is formed. Then, this transformation is inverted to give the result that is used in the computation of the stiffness matrix.

The first step is accomplished by forming a transformation from nodal displacement to strain gradient quantities using Equation 4.1. When these equations are expressed in matrix form, we have the following:

$$\begin{Bmatrix} u_i \\ v_i \end{Bmatrix} = \begin{bmatrix} 1 & 0 & -y_i & x_i & 0 & y_i/2 & x_i/2 & -y_i^2 & 0 & x_iy_i & 0 & y_i^2 \\ 0 & 1 & x_i & 0 & y_i & x_i/2 & 0 & x_iy_i & x_i^2/2 & -x_i^2/2 & y_i^2/2 & 0 \end{bmatrix} \{\varepsilon_i\}$$

$$(4.10)$$

where $\{\varepsilon_i\}$ is defined in Equation 4.5.

The transformation is formed by evaluating Equation 4.10 at each of the six nodes of the finite element. Note that this transformation is similar to the one given by Equation 3.17 which was formed in Chapter 3 to compute the equivalent continuum parameters for the hexagonal truss.

When this transformation is formed, we have the following:

$$\{d\} = [\Phi] \{\varepsilon_i\} \qquad (4.11)$$

where

$$\{d\} = [u_1 \quad u_2 \quad u_3 \quad u_4 \quad u_5 \quad u_6 \quad v_1 \quad v_2 \quad v_3 \quad v_4 \quad v_5 \quad v_6]^T$$

and

$$[\Phi] = \begin{bmatrix}
1 & 0 & -y_1 & x_1 & 0 & y_1/2 & x_1^2/2 & -y_1^2/2 & 0 & x_1y_1 & 0 & y_1^2/2 \\
1 & 0 & -y_2 & x_2 & 0 & y_2/2 & x_2^2/2 & -y_2^2/2 & 0 & x_2y_2 & 0 & y_2^2/2 \\
1 & 0 & -y_3 & x_3 & 0 & y_3/2 & x_3^2/2 & -y_3^2/2 & 0 & x_3y_3 & 0 & y_3^2/2 \\
1 & 0 & -y_4 & x_4 & 0 & y_4/2 & x_4^2/2 & -y_4^2/2 & 0 & x_4y_4 & 0 & y_4^2/2 \\
1 & 0 & -y_5 & x_5 & 0 & y_5/2 & x_5^2/2 & -y_5^2/2 & 0 & x_5y_5 & 0 & y_5^2/2 \\
1 & 0 & -y_6 & x_6 & 0 & y_6/2 & x_6^2/2 & -y_6^2/2 & 0 & x_6y_6 & 0 & y_6^2/2 \\
0 & 1 & x_1 & 0 & y_1 & x_1/2 & 0 & x_1y_1 & x_1^2/2 & -x_1^2/2 & y_1^2/2 & 0 \\
0 & 1 & x_2 & 0 & y_2 & x_2/2 & 0 & x_2y_2 & x_2^2/2 & -x_2^2/2 & y_2^2/2 & 0 \\
0 & 1 & x_3 & 0 & y_3 & x_3/2 & 0 & x_3y_3 & x_3^2/2 & -x_3^2/2 & y_3^2/2 & 0 \\
0 & 1 & x_4 & 0 & y_4 & x_4/2 & 0 & x_4y_4 & x_4^2/2 & -x_4^2/2 & y_4^2/2 & 0 \\
0 & 1 & x_5 & 0 & y_5 & x_5/2 & 0 & x_5y_5 & x_5^2/2 & -x_5^2/2 & y_5^2/2 & 0 \\
0 & 1 & x_6 & 0 & y_6 & x_6/2 & 0 & x_6y_6 & x_6^2/2 & -x_6^2/2 & y_6^2/2 & 0
\end{bmatrix}$$

The columns of the [Φ] give the displacements that identify the shape of the element when it represents the individual strain states. For example, the first column defines the displacements of the element when it is representing the rigid body motion in the x direction. The magnitude of the displacements are defined by the size of the strain gradient variable $(u_{rb})_0$.

The independent variables in the strain energy expression formed in Step 3 and given by Equation 4.6 are the strain gradient quantities that are defined in Equation 4.5 as $\{\varepsilon,\}$. As mentioned earlier, the independent variables of the strain energy expression must be expressed in terms of the nodal displacements in order to assemble the finite element stiffness matrices.

The transformation that achieves the change of variables to nodal displacements is found by inverting Equation 4.11, which gives the following:

$$\{\varepsilon,\} = [\Phi]^{-1} \{d\} \tag{4.12}$$

We are now in a position to complete the formulation of the finite element stiffness matrix. In the next step, we will transform the strain energy expression from strain gradient coordinates to nodal displacements.

A significant aside: At this point, it can be seen that the finite element and the finite difference methods have a common basis. When Equation 4.12 is inspected, we see that we have strain gradient quantities being expressed in terms of nodal displacements. Since the strain gradient coefficients are simply a specialized form of the Taylor series coefficients, the rows of Equation 4.12 can be seen to be finite difference operators (see note 4).

STEP 6—TRANSFORMATION OF STRAIN ENERGY TO NODAL DISPLACEMENT COORDINATES

The strain energy expression given by Equation 4.6 can now be transformed to nodal displacements by substituting the transformation given by Equation 4.12. When this is done, we have the following:

$$
\begin{aligned}
SE &= \frac{1}{2}\{\varepsilon,\}^T \bar{U}\{\varepsilon,\} \\
&= \frac{1}{2}\{d\}^T[\Phi]^{-T}\bar{U}\ [\Phi]^{-1}\{d\} \\
&= \frac{1}{2}\{d\}^T[K]\{d\} \tag{4.13}
\end{aligned}
$$

It should be remembered that the matrix $\bar{U}$ has already been computed in Equation 4.7 and noted that [K] is symmetric.

STEP 7—APPLICATION OF THE PRINCIPLE OF MINIMUM POTENTIAL ENERGY

The force–displacement relationship for a finite element is formed by applying the principle of minimum potential energy. The potential energy consists of two components: (1) the strain energy contained in the finite element and (2) the work done by the applied forces when the nodes of the element are displaced. Since the strain energy expression and work function are independent of each other, they will be treated separately.

The general expression for the principle of minimum potential energy for this application is the following:

$$
\frac{\partial(SE - W)}{\partial d_i} = 0; \quad i = 1,2,\ldots n \tag{4.14}
$$

We will now extract the finite element stiffness matrix from the strain energy portion of Equation 4.14. When the derivative of the strain energy expression given by Equation 4.13 is performed for a representative displacement variable, we have the following:

$$\frac{\partial\, SE}{\partial\, u_i} = \frac{1}{2}\begin{Bmatrix} 0 \\ \vdots \\ 1 \\ \vdots \\ 0 \end{Bmatrix}^T [K] \begin{Bmatrix} u_1 \\ \vdots \\ u_i \\ \vdots \\ v_n \end{Bmatrix} + \frac{1}{2}\begin{Bmatrix} u_1 \\ \vdots \\ u_i \\ \vdots \\ v_n \end{Bmatrix}^T [K] \begin{Bmatrix} 0 \\ \vdots \\ 1 \\ \vdots \\ 0 \end{Bmatrix}$$

$$= \begin{Bmatrix} 0 \\ \vdots \\ 1 \\ \vdots \\ 0 \end{Bmatrix}^T [K] \begin{Bmatrix} u_1 \\ \vdots \\ u_i \\ \vdots \\ v_n \end{Bmatrix} \qquad (4.15)$$

The two components contained in the first line of Equation 4.15 can be summed because [K] is a symmetric matrix.

When the derivatives of all n of the independent displacement variables are taken and the results of the n derivatives are summed, we have the following:

$$\frac{\partial SE}{\partial d} = [K]\{d\} \qquad (4.16)$$

The stiffness matrix being formed is equal to the [K] contained in Equation 4.16. Since we have recognized that the stiffness matrix is available without actually applying the principle of minimum potential energy, this means that the principle does not have to be applied in the computation of the stiffness matrix. The application of this principle is only required as a theoretical step to identify the stiffness matrix in the initial development.

The final force–displacement relationship for the finite element matrix is given as follows:

$$[K]\,\{d\} = \{F\} \qquad (4.17)$$

The applied force vector is found by forming a work function that depends on the form of the applied force. If the applied force consists of a point load at a node, the total force is contained in the row associated with that node. If the load is a point load that is applied at some point on the domain of the element, fractions of the load are distributed to the various nodes. However, if a distributed load is applied, the equivalent nodal loads are found by integrating a function formed by multiplying the distributed load with the appropriate displacement interpolation function.

4.5 THE SOURCE AND ROLE OF THE COMPATIBILITY EQUATION

The strain representations for a 10-node triangle are the following (Dow 2012):

$$\varepsilon_x(x,y) = (\varepsilon_x)_0 + (\varepsilon_{x,x})_0 x + (\varepsilon_{x,y})_0 y + (\varepsilon_{x,xx})_0 x^2 + (\varepsilon_{x,xy})_0 xy + (\varepsilon_{x,yy})_0 y^2$$

$$\varepsilon_y(x,y) = (\varepsilon_y)_0 + (\varepsilon_{y,x})_0 x + (\varepsilon_{y,y})_0 y + (\varepsilon_{y,xx})_0 x^2 + (\varepsilon_{y,xy})_0 xy + (\varepsilon_{y,yy})_0 y^2$$

$$\gamma_{xy}(x,y) = (\gamma_{xy})_0 + (\gamma_{xy,x})_0 x + (\gamma_{xy,y})_0 y + (\gamma_{xy,xx})_0 x^2 + (\varepsilon_{x,yy} + \varepsilon_{y,xx})_0 xy$$
$$+ (\gamma_{xy,yy})_0 y^2$$

$$(4.18)$$

As can be seen in Equation 4.18, the two normal strain models are complete second-order polynomials with the expected Taylor series coefficients. That is to say, the normal strain ε_x is expressed in terms of ε_x and its derivatives, and the normal strain ε_y is similarly represented.

However, the shear strain expression contains a term that, at first glance, seems to be an anomaly. The expected Taylor series coefficient for the xy term is $(\gamma_{xy,xy})_0$. When the expected term is equated to the coefficient that is actually present, we have the following:

$$(\gamma_{xy,xy})_0 = (\varepsilon_{x,yy} + \varepsilon_{y,xx})_0 \qquad (4.19)$$

This relationship can be recognized as the compatibility equation for planar problems. The usual explanation for this equation is that it exists as a constraint equation because the three strain components are formed from only two displacements. This is, indeed, true but the rationale behind the derivation of the equation is rarely, if ever, stated. Furthermore, there is no clear explanation of how and where this constraint equation is used (Borg 1963).

When the 10-node triangle is formed using strain gradient notation, both the derivation and the role of the compatibility equation are made clear. The 10-node triangle is formed from displacement interpolation polynomials that are complete third-order polynomials. As a result, both the u and the v interpolation polynomials contain the following terms; x^3, x^2y, xy^2, and y^3. Thus, the two polynomials have eight coefficients.

However, as seen in Equation 4.18, the three strain representations for a 10-node triangle contain nine second-order strain coefficients. Since the nine coefficients for the strains are derived from the eight coefficients of the displacement representations, there must be a term that is not independent of the other strain quantities. The coefficient of the xy term of the shear strain expression is chosen to be the term that is dependent on the eight coefficients of the displacement expressions. That is the reason that it is not the expected ($\gamma_{xy,xy}$) term and it is a function of two second-order normal strain terms.

As a result of using the physically interpretable notation, we have a clear explanation of the source and the role of the compatibility equation as it relates to the finite element method. The compatibility equation is derived in detail in both Dow (1999, 2012).

Over the years, the author has found that the validity of a mathematical model is reinforced when some salient fact concerning the problem being solved emerges from the analysis process that was not explicitly built into the model. The emergence of the compatibility equation in both two- and three-dimensional applications was an unexpected result when the strain gradient notation was first derived.

In a further extension of the notation to higher-order applications, another unexpected result is produced. When the 15-node triangle was formed, it is found that higher-order compatibility equations exist. These constraint equations consist of derivatives of the well-known compatibility equation (Dow 1999).

The identification and possible discovery of these higher-order compatibility equations once again confirms the usefulness of strain gradient notation. A survey of approximately 20 well-known elasticity, continuum mechanics, and advanced mechanics of materials books did not even hint at the existence of higher-order compatibility equations. However, one lesser known book hinted at the existence of these terms (Borg 1963).

4.6   SUMMARY AND CONCLUSIONS

The overt focus of this chapter has concerned improvements to the formulation and capabilities of stiffness matrices. We have seen that the use of the physically interpretable notation has simplified element formulation and allowed the modeling characteristics of individual elements to be identified by visual inspection.

The strain representation in elements formed with the alternative approach modeled the actual strain as well, as they could when the limited polynomial bases that are used to form an element is considered. This contrasted to the isoparametric approach which contained strain modeling errors in elements with a nonstandard shape. Mathematically, a nonstandard shape is defined as a configuration for which the Jacobian of the isoparametric transformation is not a constant. Examples of nonstandard shapes are triangles with curved edges and four-, eight-, and nine-node elements that are not parallelograms.

Although this chapter has focused on element formulation, the identification of a common basis for the finite element and the finite difference methods provides a starting point for significant improvements to error analysis and model refinement guides. If the finite element solution is treated as an approximation of a finite difference solution, quantities can be extracted from the finite element result using finite difference templates. These quantities can be used to evaluate the accuracy of the finite element solution and guide the refinement of the existing model.

For example, if the finite difference strains are compared to the finite element strains, the differences can be used to form error estimators and refinement guides. These ideas are expanded and applied in Chapters 7 and 8.

4.7   NOTES

1. The mapping that introduces the errors into the higher-order strain states is the source of the name for isoparametric elements. The mapping takes the original shape of the element onto a *regular shape*, that is, an equilateral triangle or a square of a fixed size. The coordinates of these regular shapes are the *parametric* coordinates.

    This mapping has the same form as the displacement interpolation polynomials that are used to form the stiffness matrix. The Greek prefix iso- that means *the same*. The original shape is mapped onto the regular shape with a coordinate transformation that has the same form (*iso*) as the interpolation polynomials. This allows the Gauss quadrature approximate integration scheme to be used to evaluate the large number of integrals that exist in the isoparametric formulation procedure.

2. The pedagogical complexity of the isoparametric formulation procedure can be attributed to the limited computer capabilities that existed when the finite element method was invented and developed. In order to reduce the computational effort, every element was mapped onto a predefined domain so that the integrals that must be evaluated could be computed with a simple numerical procedure known as Gauss quadrature.

3. The number of integrals that must be evaluated for a six-node iso-parametric element is found as follows. The stiffness matrix is a symmetric, (12 × 12) matrix. The total number of terms is 144. The number of off-diagonal terms is (144 − 12) = 132. Since the matrix is symmetric, the number of unique off diagonal terms is 66. The total number of unique elements is 66 + 12 = 78.

4. As noted in the main text, Equation 4.12 identifies the fact that the finite element and the finite difference methods have a common basis. This is discussed at length in Part IV of Dow (1999). As a result, new life is breathed into the finite difference method.

4.8   EXERCISES

1. This exercise forms the stiffness matrix for a three-node bar element with even spacing by applying the seven steps identified in Table 4.1.

   a. Identify the interpolation polynomial using the x-axis.

   b. Form the strain model and identify the strain states. Sketch these strain states with respect to the initial position of the bar. Hint: Put the origin at the center of the bar.

   c. Write the strain energy expression for a continuous one-dimensional bar. Then, transform the continuous expression to a finite number of strain gradient coordinates. Hint: Do not forget to include the rigid body displacement as was done in the text.

   d. Give the bar an area of 1.0 square inch and a length of 36 inches and integrate the necessary term(s) in the strain energy expression.

   e. Formulate the transformation from displacements to strain gradient terms for evenly spaced nodes. The three columns of the transformation matrix should correspond to the shape of the displacements associated with each of the three strain states. Sketch them on the initial position of the bar. Hint: First plot the horizontal displacements along the x-axis. Then plot them on the y-axis. Assume two magnitudes for the strain gradient quantities, ½ and 1.

   f. Invert the transformation formed in item e to form the transformation from strain gradient quantities to nodal displacements. Check your inversion by multiplying it by the matrix that was inverted. Hint: This is a good application for MATLAB. You can perform the inversion numerically or symbolically.

g. Assume that the bar has the following sets of displacements: 0, 0, 2; 0, 1, 0; and 1, 0.5, −1. Using the results of item f, identify the contribution of each strain state to the displacements.

h. Form the stiffness matrix for the bar. Fix one end of the bar, that is, set the displacement equal to zero. Apply a force of 100 units to the other end and find the displacements. Then, find the participation of the strain gradients. Hint: Look at item g.

2. Redo problem 1 for an element with uneven nodal spacing. Place the nodes at the following locations: 0, 16, and 36 inches.

CHAPTER 5

THE "DISTMESH_2D" MESH GENERATION PROGRAM

5.1   INTRODUCTION

Most finite element textbooks do not introduce mesh generators. As a result, there is no way to use the knowledge gained to create significant models unless some finite element package is also introduced. Typically, either these packages are so complex as to be opaque or the code is inaccessible to modification.

In this chapter, an easy-to-use and flexible mesh generator is presented so that meaningful problems can be solved immediately. When this capability is combined with the physically interpretable notation introduced in Chapter 3, new and experienced users have the capacity of immediately exploring the usefulness of the physically based notation. This capability can be used to extend the work on error estimators and mesh refinement guides presented here.

This mesh generator was chosen because it accomplishes the objective set by its authors:

> Our goal is to develop a mesh generator that can be described in a few dozen lines of MATLAB. We could offer faster implementations and refinements of the algorithm, but **our chief hope is that users will take this code as a starting point for their own work.** It is understood that the software cannot be fully state-of the-art, but it can be simple and effective and public. Persson and Strang (2004)

5.2   OBJECTIVES

This chapter has two objectives. The first is to present the mesh generation program as a stand-alone entity because of its importance in this work. The second objective is to make the MATLAB code transparent for those who want to extend the developments presented here.

The second objective is included for two reasons. First, it is always a good idea to understand the tools with which you are working. Second, the program uses MATLAB features that are far from elementary. This idea is presented directly in Persson and Strang (2004) with the following statement, "… the advanced MATLAB programming [was] contributed by the first author."

5.3   OVERVIEW OF THE MESH GENERATOR

At its heart, the MATLAB function distmesh_2d forms meshes for domains that are identified by Venn diagrams. This means that meshes can be generated for practically any configuration by combining relatively simple shapes. For example, Figure 5.1 contains meshes formed from the difference between two domains, the union of two domains, and the inter- section of two domains.

The meshes used to test the error estimators and mesh refinement guides developed in later chapters are created with this mesh generator. These error estimators and refinement guides are demonstrated with the Kirsch-type problem shown in Figure 5.1a. The domain of this problem is formed by subtracting a circle from the center of a square. The resulting finite element model is loaded in tension in the x direction. This produces well-defined stress concentrations at the top and bottom of the circum- ference of the circle (Budynas 1999).

The program that created the meshes shown in Figure 5.1 is contained here as Appendix 5A. In addition to the three meshes shown in Figure 5.1, there are two variations on the Kirsch-like problem shown in Figures 5.2a and 5.2b that are available in this program.

The first variation is formed with a nonuniform mesh. The mesh has smaller elements near the boundary of the internal circle. The second vari- ation is a square with a parabolic hole. A uniform mesh with smaller ele- ments is shown for contrast. This problem contains stress concentrations at the apexes of the parabola that are more severe than those that exist on the boundary of a circle. These problems are included as a convenience for anyone who wants to extend the work presented here. Consequently, these problems provide a good starting point for further research.

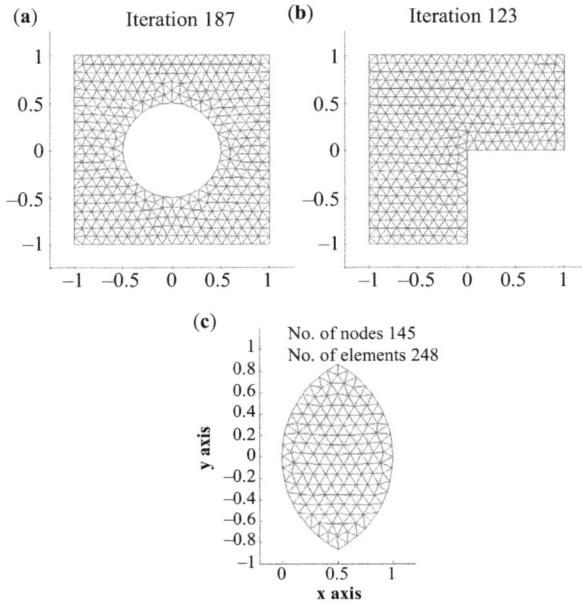

Figure 5.1. Meshes formed as Venn diagrams: (a) difference between a square and a circle, (b) union of a rectangle and a square, and (c) intersection of two circles.

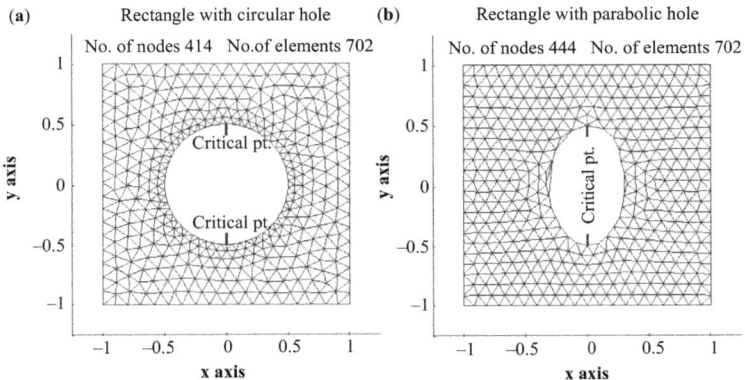

Figure 5.2. Alternate Kirsch-like problems: (a) Kirsch-like problem with a nonuniform mesh and (b) Kirsch-like problem with a parabolic hole.

This overview of the mesh generator is followed by a detailed description of its operation. Then, the operations performed by the individual lines that make up the eight steps of the distmesh_2d code are presented. The function itself is contained in Appendix 5B. The m-files for ancillary MATLAB functions used in the mesh generator are presented in Appendix 5C.

5.4 A DETAILED DESCRIPTION OF THE MESH GENERATOR

This mesh generation process begins by forming an initial mesh that can be viewed as a blank sheet of paper. A *boundary box* is filled with nodes, as shown in Figure 5.3a. Then, the nodes in the odd-numbered rows are moved to the right, as shown in Figure 5.3b. These nodes are moved so that the resulting mesh consists largely of equilateral triangles. The success of this strategy can be seen by the products of this process that are shown in Figures 5.1 and 5.2. The majority of the triangles are equilateral triangles or very close to this configuration.

In the next operation, the nodes outside of the desired configuration are removed. The key to removing these nodes is the use of a *directed distance* function. This function identifies whether a node is inside or outside the boundary. If a node is inside the boundary, the magnitude of its distance to the boundary is negative and it is retained in the mesh. If the nodal distance has a positive sign, it is removed unless it is within a small predefined distance outside the boundary. Nodes close to the boundary will be moved back to the boundary in a later operation.

After these nodes are removed, any desired fixed points are added to the mesh. The results of these two operations are shown in Figure 5.4 for the Kirsch-type problem. The four corners of the box and the two stress concentration points at the top and bottom of the circle are fixed. The six fixed points are designated with the stars in Figure 5.4b.

The nodal points that are retained are used to form a mesh with triangular elements with the Delaunay algorithm. The Delaunay triangulation

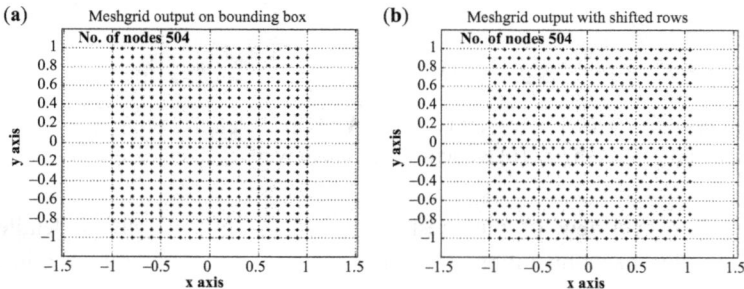

Figure 5.3. Initial mesh formulation steps: (a) original nodal locations and (b) shifted nodal locations.

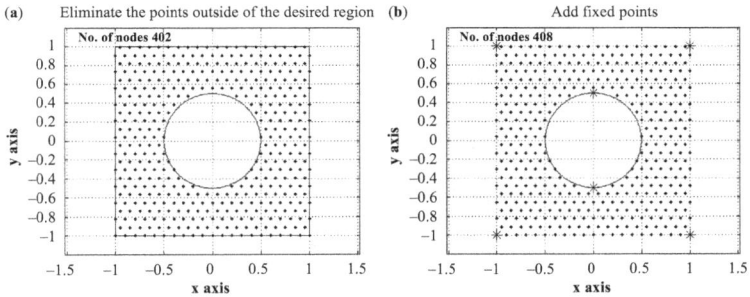

Figure 5.4. Define and fix region: (a) remove points outside of region and (b) add fixed points.

algorithm has two primary features. The first feature does not allow the nodes of adjacent triangles to exist in the smallest circumscribed circle for a triangle. This essentially does not allow the existence of any overlapping triangles. The second feature of the algorithm maximizes the minimum angle in each triangle. As can be seen in Figures 5.1 and 5.2, equilateral triangles are formed when possible by this algorithm.

However, there are two types of triangles that *must* be removed from the result of the Delaunay triangularization. The first type occurs if there is an internal hole. If there is an internal hole, the Delaunay triangulation will populate this opening using nodes that exist on the boundary. These elements must be removed or the desired configuration will not be produced.

The second type of triangles that must be removed are *degenerate triangles*. On occasion, this algorithm defines a triangle on a boundary with three points that are in a straight line. Such a triangle is defined as a degenerate triangle because its area is equal to zero.

There will often be other points between the three points that define a degenerate triangle. This nodal configuration does not violate the constraint on the intrusion of nodes of one triangle into the circumscribing circle of another triangle because the radius of the circumscribing circle for a degenerate triangle is equal to zero.

Degenerate triangles must be removed because their existence would cause havoc with a finite element stiffness matrix. The stiffness matrix would be equal to zero because a degenerate triangle has no area. Examples of degenerate triangles and their elimination will be presented later in this chapter.

As noted, the mesh generator strives to populate the domain of the problem with equilateral triangles. Equilateral triangles are desired because they have the highest quality. This concept is discussed at length in McGuire, Gallagher, and Ziemian (2000). In the case of finite elements, the quality has to do with condition number, which is related to the ratio of the maximum and minimum eigenvalues. If the condition number of an element is too large, it can negatively affect the stability of the computations.

In graphical terms, the quality of a finite element can be quantified in terms of the ratio of the inscribed circle to the circumscribed circle with the following relationship: $q = 2\, r_{inscribed}/r_{circumscribed}$. This metric varies between 1 and 0. A triangle with a q value equal to or larger than 0.5 is considered a good triangle (Field 2000). The inscribed and circumscribed circles and their quality measurement for three triangles with different configurations are shown in Figure 5.5.

As can be seen, the quality metric for an equilateral triangle is equal to 1.0, the highest possible value. The oblique triangle shown in Figure 5.5c does not quite satisfy the minimum requirement of 0.5 for an acceptable element. The quality measure for the right triangle shown in Figure 5.5b is more than acceptable.

When the initial mesh for the Kirsch-type problem shown in Figure 5.4b is inspected, it is seen that very few mesh points are actually on the circumference of the circle. The triangulation of this initial mesh is shown in Figure 5.6. The circle is not very well represented by this mesh. There are very few nodal points on the circle, and the majority of the triangles around this crude representation of the circular hole are not close to being equilateral triangles. In addition, the two ends of the square are represented by oblique triangles. Inspection of Figure 5.1a shows that both of these conditions are corrected in the final mesh, which has been iteratively improved by the mesh generator.

This mesh generator remedies these deficiencies with an iterative process that progressively improves the initial mesh by moving individual

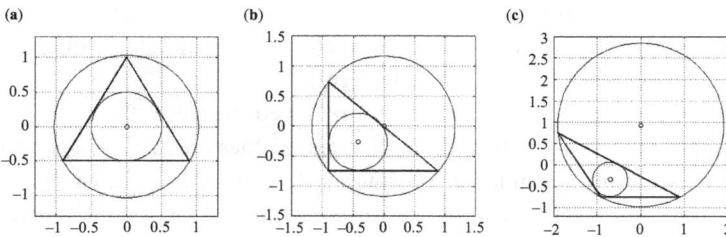

Figure 5.5. Triangles with circumscribed and inscribed circles: (a) equilateral triangle, $q = 1.0$; (b) right triangle, $q = 0.8168$; (c) oblique triangle, $q = 0.4173$.

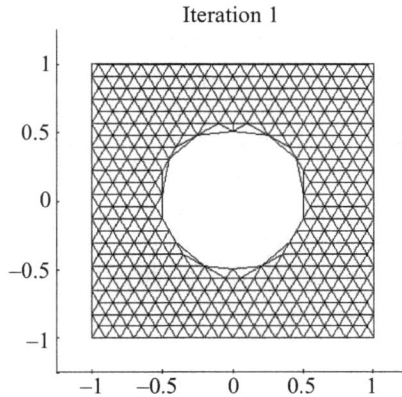

Figure 5.6. A triangulated mesh.

nodes. The *crux of the mesh generation program* is in the way in which the nodes are moved to improve the mesh.

The mesh points are moved with *two objectives*. The first is to force the distances between the nodes to have nearly equal lengths. The second is to improve the representation of the functionally defined boundaries by moving two types of points to the boundary. Points that are either close to the inside or close to the outside of the boundary are moved to the boundary.

The goal of the first objective is to produce a mesh that is composed of triangles with high-quality ratings. The reason for the second objective is obvious. The goal is to have a mesh that matches the geometry that is specified.

The effectiveness of this strategy can be seen by comparing the final mesh for the Kirsch-type problem shown in Figure 5.1a to the initial mesh shown in Figure 5.6. Since there are more nodal points on the boundary of the circular cutout, the representation of the internal circular hole in the final mesh is significantly smoother than the representation in the initial mesh. In addition, the oblique triangles on the circumference of the circle have been replaced with triangles that are close to being equilateral triangles.

The iterative process also improves the representation of the boundaries on the right and left ends of the square. There are more points on these boundaries, and the triangles on the two ends are nearer to equilateral triangles in the final mesh.

The iterative process that produces these improvements proceeds as follows. The nodes are moved by pseudo-forces that come from assuming that the edges of the triangles are linear springs or bars that can be deformed. The lengths of these *bars* are computed from the positions of the nodes that define the edges of the triangles. Then, these lengths are compared to a desired constant length. The differences between the desired lengths and the actual lengths produce loads in each of the *springs.* Only the loads that increase the separation between the nodes are retained. This is done to spread the nodes over the domain of the problem and to move interior nodes to the boundaries.

The vector sum of all of the retained forces at each node is computed and the node is moved in the direction of this vector sum by a proportion of its magnitude. If a node is moved outside of the functional boundary, it is pushed back to the boundary.

Intermediate results of this iterative process are shown in Figure 5.7. In these figures, the movements of the individual nodes for each iteration are superimposed on each other. In other words, these figures show how each node migrates as the mesh is improved. The small circles that surround the original positions of the nodes are included to give scale to the movements.

In Figure 5.7a, the movements of the nodes from their initial positions are shown after 10 iterations. As can be seen, the majority of the movement is bringing nodes to the boundaries of the domain of the problem. The continuing migration of these nodes to the boundary after 75 iterations can be seen in Figure 5.7b. Some of the nodes away from a boundary are rearranging themselves in order to produce triangles with a higher quality.

The iterative process that improves the mesh is terminated when the sum of the absolute values of the displacements of all of the points in the final iteration is below a predefined threshold. The termination criterion for the Kirsch-like problem is satisfied after 187 iterations. This contrasts to the limit of 500 iterations that is placed on the iterative process to eliminate infinite loops. This means that the termination criterion of the process was satisfied. A condition that produces an infinite loop will be discussed later.

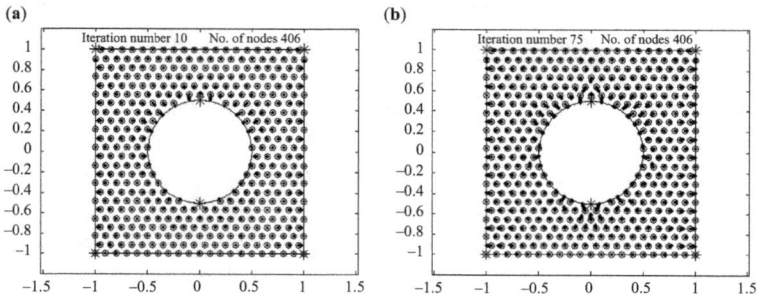

Figure 5.7. Nodal migration in the mesh generator: (a) early iteration and (b) later iteration.

The mesh generator creates the two geometric input variables needed for a finite element program, namely, the nodal coordinates and the element topology. The nodal coordinates are presented in two columns that contain the x and y coordinates of each node in the order of the nodal numbers.

An example of element topology is presented in the final three columns of Table 5.1. These three columns contain the node numbers of the three corners that identify the triangle. The first column identifies number of a triangular element for the convenience of the reader.

The four elements for which the element topology is given are highlighted in Figure 5.8. Figure 5.8a contains the element numbers for this section of the mesh. Figure 5.8b identifies the node numbers for this section

Table 5.1. Examples of element topology

Element no.	Node no.	Node no.	Node no.
⋮	⋮	⋮	⋮
459	138	155	139
460	137	155	138
461	136	137	115
462	137	138	115
⋮	⋮	⋮	⋮

(a)

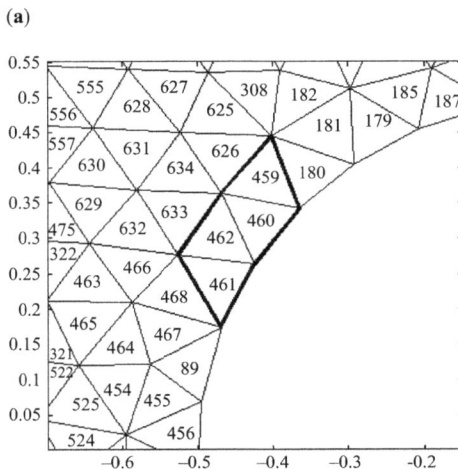

Figure 5.8. (*Continued*)

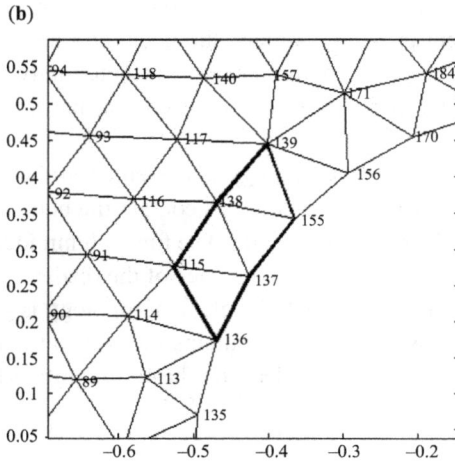

Figure 5.8. A segment of the final Kirsch-like mesh: (a) element numbering and (b) nodal numbering.

of the mesh. This figure consists of a section of the mesh that corresponds to the final mesh for the Kirsch-type problem shown in Figure 5.1a.

As advertised, this section has provided an overview of the mesh generation program. In the next section, the individual lines of the distmesh_2d MATLAB function will be presented and their operation will be described in some detail. The full program and the mesh refinement function are contained in appendixes 5A and 5B, respectively.

5.5 A LINE-BY-LINE DESCRIPTION OF THE "DISTMESH_2D" FUNCTION

The eight individual steps that make up the mesh generator distmesh_2d are described in detail in this section. The eight steps are designated in the function itself presented as Appendix 5B (Persson and Strang 2004).

This line-by-line description makes the function of each step accessible to readers, particularly those who are relatively new to MATLAB. Some of the lines of code are *elegant* or *slick* in that they accomplish several steps in one line using subtle characteristics of the operations. As a result of this advanced MATLAB programming, the objective of a specific line of code and how it is accomplished might not be obvious to anyone with a limited knowledge of MATLAB, such as the author of this book.

The eight steps are presented in detail, one step at a time. Each step is introduced with an overview of its function. Then, any inputs from the driver program are presented before the operation of the step is discussed.

STEP 1—CREATE THE INITIAL DISTRIBUTION IN THE BOUNDING BOX

This step consists of three lines of code that produces the nodal coordinates for a set of points that serve as the starting point for the final mesh. A rectangle or a square is filled with nodes in such a way that the final mesh consists of triangles that are equilateral or nearly so.

This step uses the following input from the main program:

```
Box=[-1.0,-1.0;1.0,1.0];  h0 = 0.01;
```

The quantities in the variable box identify the nodal locations of the corners of the box that contains the domain of the mesh being formed. This box can be viewed as a blank piece of paper on which the mesh is going to be created.

The variable h0 defines the nodal spacing in the x direction of the mesh points. The length of the base of the equilateral triangle is equal to h0. The spacing in the y direction is defined as part of this step such that the resulting triangles are equilateral. As we will see later, equilateral triangles are desired because of their properties.

Line 1

```
[x,y]=meshgrid(box(1,1):h0:box(2,1),box(1,2):h0*sqrt(3)/2:
box(2,2));
```

The meshgrid function forms vectors of the x and y coordinates of the points that fill the box, as shown in Figure 5.2a. The spacing of the rows in the x direction is equal to h0. The spacing in the y direction is equal to h0/sqrt(3)/2 or h0/cos(30 degrees), which is the height of an equilateral triangle that has a base that is h0 long.

Line 2

```
x(2:2:end,:) = x(2:2:end,:) + h0/2;
```

This line of code shifts the even numbered rows of nodes formed in line 1 in the positive x direction by one-half of the nodal spacing h0. This operation completes the process of positioning the nodes to form equilateral triangles. The output of this line of code is shown in Figure 5.2b. Note that the nodes on the right-hand end of the even numbered rows are outside of the bounding box.

Line 3

```
p = [x(:), y(:)];
```

The x and y coordinates of the nodal locations formed in the previous two lines of code are assembled into the variable p (for points). The number and locations of the nodal points is modified as the program proceeds. This variable, in its final form, is one of the outputs of the function distmesh_2d.

STEP 2—REMOVE POINTS OUTSIDE OF THE DESIRED GEOMETRY AND ADD FIXED POINTS

This step contains seven lines of code that removes points from those formed in the previous step. The points removed are outside of the boundary of the configuration being formed by more than a small specified distance. Then, a set of predefined fixed points are added. The number of points that remain after this step is retained throughout the analysis when uniform spacing is specified. The locations of many of these points will be moved to improve the mesh in a later step. The results of these operations are shown in Figure 5.3.

Four lines of code from the main program are used in this step. The first two lines for the Kirsch-type problem are the following inline functions:

```
fd = inline('ddiff(drectangle(p,-1,1,-1,1),dcircle(p,0,
     0,0.5))','p');
fh = inline('huniform(p)', 'p');
```

The output of an inline function can be treated like any other variable in MATLAB. The inputs to an inline function can be functions and variables. Any inline function can be replaced by a regular MATLAB function formed with an m-file. However, the use of inline functions produces a code that is less convoluted, albeit, slower.

The variable "fd" produced by the first inline function is at the heart of this mesh generator. The variable fd is the vector of the signed distances that the points "p" are away from the closest boundary of the configuration being formed. If the sign for a given point is positive, the point is outside of the boundary. If the distance to a boundary is equal to zero or negative, the point is on or inside of the boundary. This function is used in two steps in the mesh generator. In this step, it is applied to the nodal points. In the next step, it is applied to the locations of the centroids of the triangles.

In this inline function, the primary inputs are the function ddiff and the variable p. The inputs to the function ddiff are the functions drectangle and dcircle.

The function ddiff finds the distance for all points p from the boundary of a domain formed by subtracting one region from another. In the problem formed here, a circle is removed from a rectangle. The functions ddiff, drectangle, and dcircle are contained in Appendix 5D as m-files.

The variable "hd" that is defined by the second inline function is associated with forming a mesh with nonuniform spacing. This is done by removing selected points from the original set of evenly spaced points. In this work, only uniform nodal spacing is used so the procedure for forming nonuniform meshes is not discussed. The input function for this inline function huniform associates a value of 1 with each nodal point, so none of the points are removed. Nonuniform meshes are discussed in Dow (1999). The function huniform is contained in Appendix 5D.

The following two lines of code are contained in the main program and are used in this step:

```
pfix =[-1,-1;-1,1;1,-1;1,1];% This fixes the corner points
of the rectangle.
pfix=[pfix;[0.0,0.5;0.0,-0.5]];% Fix stress concentration
points on circle.
```

These two lines of code identify points that are fixed in the mesh. These points are defined according to the wishes of the user of the program. These points are not moved by the iteration process. The first set of four fixed points identifies the corners of the square portion of this configuration. The second set of two fixed points are on the circumference of the circle. They define the locations of the stress concentrations for this problem.

Line 1

```
p = p( feval ( fd, p ) < geps, : );
```

This single line of code identifies the points in the mesh that are within the domain or no more than a small distance outside of the boundary defined by the variable geps, which is specified as input. This is accomplished with three different operations in this one line of code.

The MATLAB function feval executes the function that is inside of its parentheses with the variables that are within the parentheses of feval. The variable fd identifies the shortest distance that every point p is from either the boundary of the square or the circle. These signed distances are produced by the inline function discussed earlier.

Then, the logical operation < is performed to determine if the point is within the domain or no more than the variable geps outside of the boundary. If the point satisfies this condition, a logical variable is set to 1 otherwise it is set to zero. Finally, if the logical variable is equal to 1, the point p is retained. That is to say, all of the points inside or very close to the boundaries of the problem are retained. Any of these points that are outside of the boundary are pushed back to the boundary in a later step. The output of this operation is presented in Figure 5.3a.

Lines 2 and 3

```
r0 = 1 ./ feval ( fh, p ).^2;
p = [ pfix; p(rand(size(p,1),1) < r0 ./ max ( r0 ),: ) ];
```

These lines of code have the capacity to perform several operations if a nonuniform mesh is defined. However, since only uniform meshes will be used in this work, the procedure for forming nonuniform meshes will not be discussed. In this work, the inline function fh defines the mesh as being uniform.

Since only uniform meshes are used, these lines of code perform only one function. The fixed nodes are added to the top of the list of nodes in this step. The result of this operation is seen in Figure 5.3b.

Lines 4, 5, and 6

```
[ q, i, j ] = unique ( p, 'rows', 'first' );
k = unique ( i );
p = p(k,:);
```

These lines of code are described by a comment in the version of distmesh_2d being used here as follows:

```
% The above obscure commands are written so that:
% * We ALWAYS keep the fixed points at the beginning of
   the P array.
%     * We remove any points which are duplicates of
        these points.
```

```
%        That way, we do not later allow the fixed points
         to move,
%          because we know they are at the beginning of
           the array,
%            and DELAUNAYN stops complaining about dupli-
           cate points.
% JVB (John Burkardt), 09 June 2012.
```

Line 7

```
N = size ( p, 1 );
```

This line of code identifies the number of nodes that remain in the mesh. It is used in later operations.

5.5.1 ITERATION PORTION OF THE FUNCTION

We have finished with the initialization of the mesh generation.

We now enter the iterative portion of the program with the following line of code:

```
while ( iteration < iteration_max )
```

This line of code stops the program after a predefined number of iterations are reached. In this application, the variable iteration_max is set to 500. This is to stop any possibility of an infinite loop. The desired termination of the mesh generator occurs in step 8 when the sum of the movements of the interior nodes is less than a predefined value, dptol. As mentioned earlier, the final mesh shown in Figure 5.1a is the result of 187 iterations.

STEP 3—TRIANGULATION BY THE DELAUNAY ALGORITHM

This step contains six lines of code that identifies the topology of the triangles that are formed from the existing set of nodal points. Each triangle is identified by a row that contains the node numbers for the three nodes that define its vertices. Samples of this output are contained in the last three lines of Table 5.1.

The following operations are performed in this step. The current nodal points are saved for comparison with the next set of nodal locations in order to decide if any large movements have taken place. If there is any large movement of the nodes, the mesh is retriangulated. The centroids of the triangles are then found in order to see if they are outside of the boundary or if they are *inside* of the boundary by a given small amount. If so, these triangles are eliminated. The first group is eliminated because they are outside of the boundary. The second group is eliminated because they are degenerate triangles (three points in a straight line) or very near to being such. Although triangles are eliminated, none of the nodal points are removed.

Line 1

```
if ( ttol < max ( sqrt ( sum ( ( p - pold ).^2, 2 ) ) / h0 ) )
```

This line determines if there is enough movement of the nodes to warrant a retriangulation. The absolute value of the movement of all of the nodes is scaled to the size of the initial mesh spacing. All of these values are summed. Finally, the sum is tested to see if it exceeds a threshold value. If the total movement exceeds a threshold level, retriangulation occurs.

Line 2

```
pold = p;
```

This line of code saves the current nodal locations. These values are compared to the next set of displacements with the computations contained in line 1 in the next iteration.

Line 3

```
t = delaunayn ( p );
```

This line of code triangulates the current set of nodal points. The output variable t contains the topology of the individual triangles, that is, a row of three node numbers that define the triangle. As mentioned earlier, this function attempts to mesh the domain with equilateral triangles that do not overlap.

Line 4

```
triangulation_count = triangulation_count + 1;
```

This is a counter for identifying the number of times the mesh has been triangulated. This variable is initialized as zero in the main program.

Line 5

```
pmid = ( p(t(:,1),:) + p(t(:,2),:) + p(t(:,3),:) ) / 3;
```

This line of code computes the centroid of each of the triangles. These values are used in the next line of code to determine if a triangle is to be eliminated.

The centroid of a triangle is computed by summing the coordinates of the three nodal points that identify the triangle and dividing by three. For example, the x component of the centroid of element 460 in Table 5.1 is computed with the following relationship:

$$(p(137,1) + p(155,1) + p(138,1))/3.$$

The first step in this calculation is to identify the nodal topology for each element. This information is inserted into the nodal location vector contained in the variable p. However, the element topology is contained in the variable t. So, instead of *directly* inserting the element topology into the vector of nodal locations as is done in the previous paragraph, the element topology is inserted *indirectly* by referring to the location where it resides. That is to say, the following operation in line 5 p(t(460,1)) is identical to p(137,1).

In computer coding terms, the locations of the centroids are found using indirect addressing. The nodal locations are inserted indirectly. Instead of inserting the nodal numbers directly into the coordinate locations, the variable that contains these values is substituted into the variable that contains the nodal locations.

Line 6

```
t = t( feval ( fd, pmid ) < -geps, : );
```

This line of code is somewhat analogous to line 1 in step 2. Line 1 in step 2 eliminates nodes that are outside of the domain of the problem by a small distance defined by the input variable, geps. In contrast, while eliminating triangles that are outside of the boundary, this line of code also eliminates triangles that are inside of the boundary by the small distance specified by the variable, geps.

Figure 5.9 identifies the centroids of the 34 triangles that are removed from the finite element model after the fourth iteration of the mesh generator. The centroids of the triangles to be eliminated are denoted by circles.

An example of a triangle with its centroid outside of the boundary is shown at the top of the circle in Figure 5.9. The three nodes that define the

Centroids of triangles removed

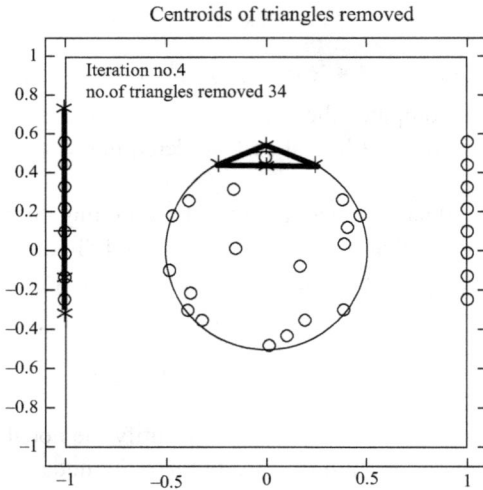

Figure 5.9. Example triangles removed by line 7.

triangle are indicated with stars. The circle with the cross inside of it identifies the centroid of this triangle. The triangle to be removed is identified with the three bold lines.

An example of a triangle that is not outside of the boundary of the domain is shown on the left-hand boundary of the square with a bold straight line in Figure 5.9. This example consists of a degenerate triangle—a triangle with zero area formed by three nodes in a straight line.

The stars at the ends of this line indicate the locations of two of the nodes that make up this degenerate triangle. The third node is designated by the star near the bottom of the line. The centroid of this triangle is identified by the circle with the cross in it located at approximately $y = 0.1$.

A key point to note is that there are centroids for other degenerate triangles contained on the interior of this degenerate triangle. This implies that the radius of the circumscribed circle for a degenerate triangle is equal to zero since the Delaunay algorithm does not allow the nodes of other triangles to be inside of the circumscribed circle of a triangle. It is important to note that the number of nodes in this mesh is unaffected by the removal of these triangles.

STEP 4—IDENTIFIES THE THREE EDGES OF EACH TRIANGLE

This step consists of two lines of code that *uniquely* identifies the edges of each triangle in a column vector that contains the node numbers of the two nodes that define the edge. In other words, this step identifies the topology of the edges of the triangles.

This information is used in step 6 to find the lengths of the edges. By treating these edges as deformed bars, these lengths are used to find forces that move the nodes apart in order to improve the mesh. The forces are created by comparing the actual lengths of the bars to a desired length for the bars.

Line 1

```
bars = [ t(:,[1,2]); t(:,[1,3]); t(:,[2,3]) ];
```

This line of code identifies the nodal pairs for each edge of each of the triangles. For example, when the expression t(:, [1,2]) is applied to Table 5.1, the result is [138 and 155], [137, 155], [136, 137], and [137,138]. That is to say, this is the nodal topology of the edges.

Each edge, except those on the boundary, will be contained in this vector two times. This is the case because each interior edge is associated with two triangles. For example, two triangles might have nodes 5 and 6 in common. As a result, this pairing would be present in the output of this step twice. Note that the variable that designates these pairs of nodes is bars.

Line 2

```
bars = unique ( sort ( bars, 2 ), 'rows' );
```

This line of code performs three operations. The *sort* operation puts the smallest nodal number for a given edge or bar as the first element in each row. For example, an interior edge for two adjacent triangles could be defined in the output of line 1 as node 6 followed by node 5 in two places. The order would be changed by the sort operation to put the reference to node 5 first in both cases. As a result, there would be two identical entries for this edge, namely, [5 6].

The *unique* operation performs two functions. It rearranges the entries into descending order and eliminates repeated rows. As a result, only one entry for each edge is defined.

The MATLAB operations for forming this list are straightforward. However, the use of this list in step 6 is not straightforward. It utilizes a characteristic of the MATLAB function *sparse* that is not obvious.

The physical interpretation of the output of this step is significant when it is used in step 6. This interpretation will be identified with the rows associated with nodes 1 and 7, which are the following:

$$\text{bars} = \begin{matrix} 1\ 7 \\ 1\ 8 \\ 1\ 30 \\ \cdot \\ \cdot \\ \cdot \\ 7\ 8 \\ 7\ 30 \\ 7\ 31 \\ 7\ 32 \end{matrix}$$

As can be seen, node 1 is connected to nodes 7, 8, and 30. This means that node 1 connects with three other nodes. When the list for node 7 is viewed, it appears that node 7 is connected to four nodes, namely, nodes 8, 30, 31, and 32. However, this list is not complete. As can be seen in the entries for node 1, node 7 is also connected to node 1. Thus, node 7 is actually connected to five nodes.

As we will see in step 6, the MATLAB function *sparse* recognizes this fact.

STEP 5—CREATES THE GRAPHICAL OUTPUT FOR THIS MESH

This step plots the mesh of triangles for this iteration of the mesh generation procedure.

```
trimesh ( t, p(:,1), p(:,2), zeros(N,1), 'EdgeColor', 'k',
'Linewidth', 1)
view(2), axis equal, axis([-1.25 1.25 -1.25 1.25]),
drawnow, grid off
```

These are standard MATLAB commands. The topology of the triangles is contained in the rows of the t matrix. The x and y coordinates are contained in the two columns of the p matrix.

STEP 6—MOVES THE MESH POINTS BASED ON BAR LENGTHS AND FORCES

This step is the crux of this program. It generates the movement of the nodes with the goal of improving the mesh. This step moves nodes apart and, hence, toward the boundaries while attempting to form equilateral triangles. This step contains nine lines of code.

These goals are pursued by treating the edges of the triangles as bars or linear springs. If the length of the element varies from a desired length, a force is generated on the nodal points at the two ends of the bar.

This desired length is equal to a scaled value computed in line 4 of this step. Only the forces that repel the nodes from each other are retained so that the nodes expand toward the borders. The repelling forces at the individual nodes are summed and the nodes are moved in the direction of the force. If a node is taken outside of the boundary, it is forced back to the boundary.

Any force that exists on a fixed point is set to zero so the point will not move.

Line 1

```
barvec = p(bars(:,1),:) - p(bars(:,2),:);
```

This line forms the vector components of the individual bar lengths by subtracting the nodal locations of one end of the bar from the other end of the bar.

Line 2

```
L = sqrt ( sum ( barvec.^2, 2 ) )
```

This line forms the actual length of the bars by first squaring the vector components of each bar length given by barvec. These two components are then added and the square root is taken to give the scalar length of each bar.

Line 3

```
hbars = feval ( fh, (p(bars(:,1),:)+p(bars(:,2),:))/2 );
```

As it is being used here, the mesh generation program is attempting to create as uniform of a mesh as it can, that is, a mesh with elements of nearly equal shapes and sizes. The function "fh" can be used to create a mesh that is not uniform. However, in this work, only uniform meshes are used. Consequently, the function fh in this case is the function *uniform* that generates a column vector of ones that has the same number of elements as there are bars. That is to say, in this case, line 3 could be replaced by hbars = ones(size(bars,1).

Line 4

```
L0 = hbars * Fscale * sqrt ( sum(L.^2) / sum(hbars.^2) )
```

This line is forming vectors of desired lengths for each of the bars. In this case, these lengths will be uniform. This desired length consists of three multiplicative components: (1) The value for hbar formed in the previous step, (2) a program defined scale factor of 1.2, and (3) a scale factor based on the ratios of the actual length of the bars and hbar, which for the first iteration is equal to 0.1449. This contrasts to the initial nodal spacing contained in the input of 0.1.

Line 5

```
F = max ( L0 - L, 0 )
```

This step forms a vector that acts as the magnitudes of pseudo-forces that move points in the process of improving the mesh. It consists of the difference between the desired length and the actual length of the bar.

Line 6

```
Fvec = F ./ L * [1,1] .* barvec;
```

This step creates the vector form of the pseudo-force. This is accomplished by multiplying the magnitude of the pseudo-force by the unit vector of the actual length of the bar in a three-step process. The force is first divided by the magnitude of the bar length. Then it is multiplied by the vector [1, 1] in order to form two components of equal size. Then it is multiplied by the components of the length of the bar.

Line 7

```
Ftot=full(sparse(bars(:,[1,1,2,2]),ones(size(F))*[1,2,
1,2],[Fvec,-Fvec],N,2))
```

This line uses the sparse operator to sum all of the forces that are applied to the individual nodes of the mesh.

Line 8

```
Ftot(1:size(pfix,1),:) = 0
```

This line sets the forces acting on the fixed points to zero.

Line 9

```
p = p + deltat * Ftot;
```

This line adds the movement of the nodes due to the repelling forces to the position the nodes had on entering this iteration. In other words, the variable p is the new x and y coordinates of the individual points.

STEP 7—BRINGS OUTSIDE POINTS BACK TO THE BOUNDARY

The five lines of code of this step bring points outside of the boundary back to the boundary.

Line 1

```
d = feval ( fd, p );
```

This line forms a vector of the magnitudes of the distances that the individual points are away from a boundary. If the distance is positive, the point is outside of the boundary. The inline function fd defined in step 2 identifies the geometry of the problem.

Line 2

```
ix = d > 0;
```

This line identifies the points that are outside of the boundary with a logical indicator. That is to say, if the value of the variable ix associated with a node is equal to 1, the point is outside of the boundary. If the value is 0, the point is either inside the boundary or on it.

Line 3

```
dgradx = ( feval(fd, [p(ix,1)+deps,p(ix,2)]) - d(ix) ) / deps;
```

This line forms the x component of the distance that a point is outside of the boundary.

Line 4

```
dgrady = ( feval(fd, [p(ix,1),p(ix,2)+deps]) - d(ix) ) / deps;
```

This line forms the y component of the distance that a point is outside of the boundary.

Line 5

```
p(ix,:) = p(ix,:) - [ d(ix) .* dgradx, d(ix) .* dgrady ];
```
This line moves the points outside of the boundary back to the boundary.

STEP 8—TERMINATES THE PROCESS

If the sum of the movements of all of the points is less than a predefined constant, terminate the mesh refinement process.

Line 1

```
if (max(sqrt( sum ( deltat * Ftot ( d < -geps,:).^2, 2 ) )
/ h0 )< dptol)
```
This line computes the total movement of all of the points and compares it to a predefined value. If the computed value exceeds the termination value, the next iteration is started.

```
        break;
    end
 end
```


5.6   SUMMARY AND CONCLUSION

This chapter has provided a detailed description of the operation of this mesh generation program on three levels. First, an overview is presented that identifies the way in which the domains are defined. Second, a detailed description is given of the process by which the initial mesh is formed and then improved until it is deemed adequate for use. Finally, the individual lines of the distmesh_2d MATLAB function that performs these operations are described. These descriptions are augmented with figures that present meshes at various stages in the formulation process.

These descriptions are presented in order to make the operation of the mesh generator transparent. This allows the user to better understand its capabilities and limitations. For example, the mesh with the circular hole shown in Figure 5.1a provides a satisfactory representation of the problem. It is used in succeeding chapters in the development of the error estimators and refinement guides.

This mesh was produced after 187 iterations. In other words, it satisfied the prescribed termination criterion of a limited amount of movement of the nodes as was described in step 8. The program currently has a limitation of 500 iterations in order to eliminate the possibility of an infinite loop.

In contrast, the domain with the parabolic hole shown in Figure 5.2b did not satisfy the termination criterion. It stopped after the arbitrary limit of 500 iterations was reached. As a result, this mesh is suspect. It should not be accepted without close scrutiny. When a section of this mesh is magnified as shown in Figure 5.10a, we see a reason for concern. Not all of the nodal points near the boundary are on the boundary.

In order to identify a reason for the lack of convergence, a close-up of the same region after 499 iterations is shown in Figure 5.10b. When the two figures are compared, we see the reason for the failure to converge. The nodes in iteration 499 are actually closer to the boundary than they are in the iteration 500. Thus, in this small region, there is significant movement. As a result, the termination criterion would not be satisfied because the position of the nodes would oscillate. Without the limitation on the number of iterations, it seems that the attempt to improve this mesh would continue ad infinitum because the termination criterion is never satisfied.

In order to gather some insight into the behavior of the mesh generator, two variables that are involved with the movement of nodes and elimination of triangles were modified. This was done to see the effects on the final mesh and on the iterative process. These meshes that are the results of 500 iterations are shown in Figure 5.11. Thus, neither of these changes led to convergence.

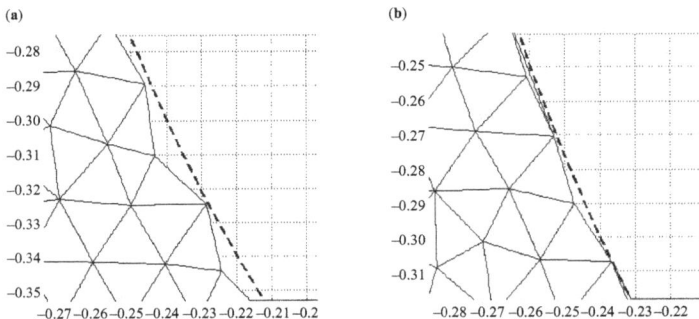

Figure 5.10. Nonconvergence due to mesh oscillation: (a) mesh = 0.02, geps = 0.2h0, 500 iterations and (b) mesh = 0.02, geps = 0.2h0, 499 iterations.

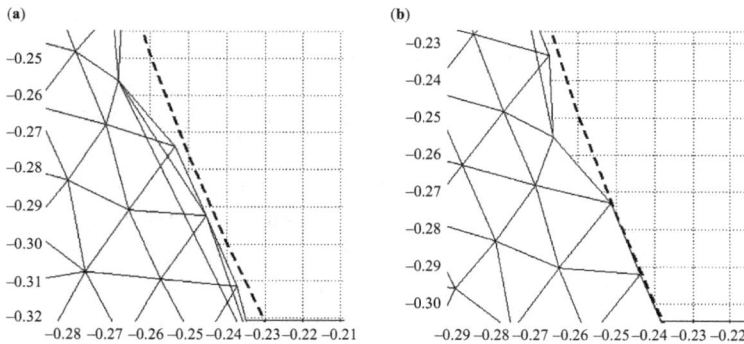

Figure 5.11. Flawed meshes: (a) mesh = 0.2, geps = 0.00001h and (b) mesh = 0.2, geps = 0.4, Fscale = 1.45.

In Figure 5.11a, the variable geps was reduced from 0.2h0 to 0.00001h0 as an experiment. This would reduce the distance for which nodes outside of the boundary would be removed and for which triangles inside of the boundary would be removed. As can be seen, the final mesh is unsatisfactory for two reasons. The tentative boundary nodes did not make it to the boundary and there are triangles superimposed on each other.

In Figure 5.11b, the variable geps was increased from 0.2h0 to 0.4h0 and the scale factor for the desired length of the bar elements in step 6 was increased from 1.2 to 1.45. As can be seen, the nodes got closer to the boundary, but overlapping triangles still exist.

The increase in geps was designed to eliminate the overlapping triangles. As can be seen, this was not completely successful because one overlapping triangle still exists. The increase in the scale factor was designed to increase the movement of the nodes toward the boundary. This seems to have been partially successful. However, the mesh still has nodes away from the boundary. This mesh is unacceptable.

The purpose of presenting these failed meshes is to reduce complacency and to reinforce the reason for describing the mesh generator is such detail. With the understanding given by these descriptions, the diagnoses and possible strategies for forming adequate meshes are available.

The two most obvious are to experiment with the initial spacing and with the placing of fixed points on the boundary of the parabola. However, the best strategy would be to investigate the Delaunay algorithm for triangulating the mesh to determine its behavior. Then, the mesh could be tailored to improve its operation.

In conclusion, this mesh generator delivers what it promises. It is simple and accessible. It provides a successful starting point for the work presented here. It should be noted that there is a discussion of improvements that could be made to the mesh generator in Dow (1999). Furthermore, Dow (1999) indicates that this mesh generator is also applicable to the following applications: (1) moving boundaries, (2) meshes for where the boundaries are defined by images, and (3) three-dimensional problems.

5.7 EXERCISES

1. Use the MATLAB code contained in the appendixes to run one of the other problems that are available. For example, find the final mesh for the Kirsch-like problem with the variable mesh. Identify the number of nodes and the number of elements in the final mesh. How many iterations did it take to get this result?
2. Redo the case that was chosen for Exercise 1 with a finer mesh size. Identify the number of nodes and the number of elements in the final mesh. How many iterations did it take to get this result?
3. Redo the problem that you are using and insert some fixed points. These points can be on a boundary or on the interior of the domain. Observe how the mesh is changed. This problem is designed to show how the user can change the results of the mesh. This is a precursor to how the error estimators are going to be used by the mesh refinement guide to introduce additional elements into regions of high stress concentration.

APPENDIX 5A MATLAB CODE FOR MESH GENERATION DRIVER

5A.1 INTRODUCTION

This appendix presents an annotated m-file for the driver program for the mesh generator distmesh_2d. This is the program that formed all of the figures contained in the main body of this chapter. The program text with the numbering came from inserting line 11 as a standalone entry in the Command Window of the MATLAB program as it is saved in the computer.

This program creates six different meshes as identified in lines 15 through 20. As it stands now, the variable Mesh is set to 1. This case will create a Kirsch-like problem mesh shown in Figure 5.1a in the main text. By engaging the values for the variable Mesh of 2 to 6, the meshes shown in Figures 5.1b, 5.1c, 5.2a and 5.2b are created.

In the succeeding chapters, other capabilities will be added. In Chapter 6, the capability of forming finite element models for the Kirsch-like problem is presented. In Chapters 8 and 9, the capabilities of adding error estimators and then mesh refinement will be added.

5A.2 MATLAB CODE

```
1. clc
2. clear all
3. close all
4. format compact
5. %
6. % This program forms meshes for six problems using the
   mesh generation
7. % program Distmesh. At its heart, this program forms
   meshes for meshes that
8. % are identified with Venn diagrams. This program demon-
   strates the difference,
9. % union and intersection operations.
10. %
11. %dbtype C:\DistMeshMain.m%This puts line numbers on
    this file.
12. %%
13. % Identify the problem for which to form mesh by eliminat-
    ing the % from on of
14. % the cases, e.g., to form the mesh for the square with a
    parabolic hole, Mesh=3.
15. Mesh=1 % square with circular hole, uniform mesh.
16. % Mesh=2 % Square with circular hole, mesh refined at cir-
    cle boundary.
17. % Mesh=3 % Square with parabolic hole, uniform mesh.
18. % Mesh=4 % Symmetric 1/4 of rectangle with circular hole,
    uniform mesh.
19. % Mesh=5 % Rectangle plus rectangle, uniform mesh.
20. % Mesh=6 % Intersecting circles
21. %
22. if (Mesh==1)
23.    % Start - Rectangle with circular hole, uniform mesh:
24.    %          Formed with difference operation.
25.    %
```

```
26.  box=[-1.0,-1.0;1.0,1.0]; % This defines the box in which the
     mesh is placed.
27.  fd=inline('ddiff(drectangle(p,-1,1,-1,1),dcircle(p,0,0,0.5))','p');
28.  pfix=[-1,-1;-1,1;1,-1;1,1];% This fixes the corner points of the
     rectangle.
29.  pfix=[pfix;[0.0,0.5;0.0,-0.5]];% Fix critical pts on circle.
30.  %meshsize=0.05
31.  meshsize=0.1
32.  %[p,t]=distmesh_2d(fd,@huniform,meshsize,[-1,-1;1,1],500,
     pfix);
33.  [p,t] = distmesh_2d(fd,@huniform,meshsize,box,500,pfix);
34.  A=size(p); NoNodes=A(1,1)
35.  B=size(t); NoElements=B(1,1)
36.  figure(1)
37.  hold on
38.  plot([0.0,0.0],[+0.5,+0.4],'k','linewidth',2)
39.  text(-.2,.35,'Critical Pt.','fontweight','bold')
40.  plot([0.0,0.0],[-0.5,-0.4],'k','linewidth',2)
41.  text(-.2,-.35,'Critical Pt.','fontweight','bold')
42.  title('Rectangle with Circular Hole','fontsize',14)
43.  xlabel('X Axis','fontsize',14)
44.  ylabel('Y Axis','fontsize',14)
45.  % Insert number of nodes and elements on figure.
46.  FigText2=sprintf('No. of Nodes %d',NoNodes);
47.  text(-1.0,1.1,FigText2,'fontsize',10,'fontweight','bold')
48.  FigText2=sprintf('No. of Elements %d',NoElements);
49.  text(0.1,1.1,FigText2,'fontsize',10,'fontweight','bold')
50.  % End - Rectangle with circular hole, uniform mesh:
51. end
52. %
53. if (Mesh==2)
54.  % Start - Rectangle with circular hole, refined at circle bound-
     ary:
55.  %        Formed with difference operation.
56.  fd = inline('ddiff(drectangle(p,-1,1,-1,1),dcircle(p,0,0,0.5))',
     'p');
57.  fh = inline('min(4*sqrt(sum(p.^2,2))-1,2)','p');
58.  pfix=[-1,-1;-1,1;1,-1;1,1];% This fixes the corner points of the
     rectangle.
59.  pfix=[pfix;[0.0,0.5;0.0,-0.5]];% Fix critical pts on circle.
60.  %meshsize=0.05
```

```
61.   meshsize=0.055
62.   [p,t] = distmesh_2d(fd,fh,meshsize,[-1,-1;1,1],500,pfix);
63.   A=size(p); NoNodes=A(1,1)
64.   B=size(t); NoElements=B(1,1)
65.   figure(1)
66.   hold on
67.   plot([0.0,0.0],[+0.5,+0.4],'k','linewidth',2)
68.   text(-.2,.35,'Critical Pt.','fontweight','bold')
69.   plot([0.0,0.0],[-0.5,-0.4],'k','linewidth',2)
70.   text(-.2,-.35,'Critical Pt.','fontweight','bold')
71.   title('Rectangle with Circular Hole','fontsize',14)
72.   xlabel('X Axis','fontsize',14)
73.   ylabel('Y Axis','fontsize',14)
74.   % Insert number of nodes and elements on figure.
75.   FigText2=sprintf('No. of Nodes %d',NoNodes);
76.   text(-1.0,1.1,FigText2,'fontsize',10,'fontweight','bold')
77.   FigText2=sprintf('No. of Elements %d',NoElements);
78.   text(0.1,1.1,FigText2,'fontsize',10,'fontweight','bold')
79.   % End - Rectangle with circular hole, refined at circle boundary:
80. end
81. %
82. if (Mesh==3)
83.   % Start - Rectangle with parabolic hole, uniform mesh:
84.   % The major and minor axes are defined in dparabola as 0.3
      and 0.5.
85.   % Formed with difference operation.
86.   fd       =        inline('ddiff(drectangle(p,-1,1,-1,1),dparabo-
      la(p,0,0))','p');
87.   pfix=[-1,-1;-1,1;1,-1;1,1];% This fixes the corner points of the
      rectangle.
88.   pfix=[pfix;[0.0,0.5;0.0,-0.5]]% Fix critical pts on parabola.
89.   meshsize=0.05
90.   %meshsize=0.1
91.   [p,t] = distmesh_2d(fd,@huniform,meshsize,[-1,-1;1,1],500,
      pfix);
92.   A=size(p); NoNodes=A(1,1)
93.   B=size(t); NoElements=B(1,1)
94.   figure(1)
95.   hold on
96.   plot([0.0,0.0],[+0.5,+0.4],'k','linewidth',2)
```

```
97.    plot([0.0,0.0],[-0.5,-0.4],'k','linewidth',2)
98.    text(0.0,-.3,'Critical Points.','rotation',90,'fontweight','bold')
99.    title('Rectangle with Parabolic Hole','fontsize',14)
100.   xlabel('X Axis','fontsize',14)
101.   ylabel('Y Axis','fontsize',14)% Insert number of nodes and
       elements on figure.
102.   FigText2=sprintf('No. of Nodes %d',NoNodes);
103.   text(-1.0,1.1,FigText2,'fontsize',10,'fontweight','bold')
104.   FigText2=sprintf('No. of Elements %d',NoElements);
105.   text(0.1,1.1,FigText2,'fontsize',10,'fontweight','bold')
106.   % End - Rectangle with parabolic hole, uniform mesh.
107. end
108. %
109. if (Mesh==4)
110.   % Start - Rectangle with circular hole, uniform mesh, 1/4 prob-
       lem:
111.   % Formed with difference operation.
112.   fd = inline('ddiff(drectangle(p,-1,1,-1,1),dcircle(p,-1,-1,0.5))',
       'p');
113.   pfix=[-1,-1;-1,1;1,-1;1,1];% This fixes the corner points of the
       box.
114.   pfix=[pfix;[-1.0,-0.5;-0.5,-1.0]];% Fix corners on circle.
115.   %meshsize=0.05
116.   meshsize=0.1
117.   [p,t] = distmesh_2d(fd,@huniform,meshsize,[-1,-1;1,1],500,p-
       fix);
118.   A=size(p); NoNodes=A(1,1)
119.   B=size(t); NoElements=B(1,1)
120.   figure(1)
121.   hold on
122.   plot([-1.0,-1.0],[-0.5,-0.62],'k','linewidth',2)
123.   text(-1.15,-0.7,'Critical Pt.','fontweight','bold')
124.   title('Rectangle with Circular Hole, 1/4 Problem','fontsize',14)
125.   xlabel('X Axis','fontsize',14)
126.   ylabel('Y Axis','fontsize',14)
127.   % Insert number of nodes and elements on figure.
128.   FigText2=sprintf('No. of Nodes %d',NoNodes);
129.   text(-1.0,1.1,FigText2,'fontsize',10,'fontweight','bold')
130.   FigText2=sprintf('No. of Elements %d',NoElements);
131.   text(0.1,1.1,FigText2,'fontsize',10,'fontweight','bold')
```

132. % End - Rectangle with circular hole, uniform mesh, 1/4 problem:

133. end

134. %

135. if (Mesh==5)

136. % Start - Rectangle plus rectangle, uniform mesh:

137. % Ninety degree corner formed with union operation.

138. fd = inline('dunion(drectangle(p,-1,0,-1,1),drectangle(p,0,1, 0,1))','p');

139. pfix=[-1,-1;-1,1;0,1;1,1;1,0;0,0;0,-1];% This fixes the corner points.

140. %meshsize=0.05

141. meshsize=0.1

142. [p,t] = distmesh_2d(fd,@huniform,meshsize,[-1,-1;1,1],500,p-fix);

143. A=size(p); NoNodes=A(1,1) % Number of nodes for figure.

144. B=size(t); NoElements=B(1,1) % Number of triangles for figure.

145. figure(1)

146. hold on

147. plot([0.0,0.2],[0.0,-0.2],'k','linewidth',2)

148. text(0.1,-0.3,'Critical Point.','fontweight','bold')

149. title('Ninety Degree Corner','fontsize',14)

150. xlabel('X Axis','fontsize',14)

151. ylabel('Y Axis','fontsize',14)

152. % Insert number of nodes and elements on figure.

153. FigText2=sprintf('No. of Nodes %d',NoNodes);

154. text(-1.0,1.1,FigText2,'fontsize',10,'fontweight','bold')

155. FigText2=sprintf('No. of Elements %d',NoElements);

156. text(0.1,1.1,FigText2,'fontsize',10,'fontweight','bold')

157. % End - Rectangle plus rectangle, uniform mesh:

158. end

159. %

160. if (Mesh==6)

161. % Start - Two circles intersecting:

162. % Formed with intersection operation.

163. fd = inline('dintersection(dcircle(p,0,0,1),dcircle(p,1,0,1))', 'p');

164. pfix=[0,0;1,0]; % Fixes the points on far left and right.

165. %meshsize=0.05

166. meshsize=0.1

167. [p,t] = distmesh_2d(fd,@huniform,meshsize,[-1,-1;1,1],500, pfix);
168. A=size(p); NoNodes=A(1,1) % Number of nodes for figure.
169. B=size(t); NoElements=B(1,1) % Number of triangles for figure.
170. figure(1)
171. hold on
172. axis([-0.2,1.2 -1.0 1.2])
173. %plot([0.0,0.2],[0.0,-0.2],'k','linewidth',2)
174. title('Intersecting Circles','fontsize',14)
175. xlabel('X Axis','fontsize',14)
176. ylabel('Y Axis','fontsize',14)
177. % Insert number of nodes and elements on figure.
178. FigText2=sprintf('No. of Nodes %d',NoNodes);
179. text(0.1,1.1,FigText2,'fontsize',10,'fontweight','bold')
180. FigText2=sprintf('No. of Elements %d',NoElements);
181. text(0.1,0.95,FigText2,'fontsize',10,'fontweight','bold')
182. % End - Two circles intersecting:
183. end

APPENDIX 5B MATLAB CODE FOR THE FUNCTION DISTMESH_2D

5B.1 INTRODUCTION

This appendix presents an annotated m-file for the mesh generator dist-mesh_2d. It is discussed in detail in the main text and its source is Persson and Strang (2004) in the main text.

5B.2 MATLAB CODE

1. function [p, t] = distmesh_2d (fd, fh, h0, box, iteration_max, pfix, ...
2. varargin)
3.
4. %***
****

```
 5. %
 6. %dbtype C:\ZZZDistmesh\distmesh_2d.m %This lists program
    with line nos.
 7. %% DISTMESH_2D is a 2D mesh generator using distance
    functions.
 8. %
 9. % Licensing:
10. %      (C) 2004 Per-Olof Persson.
11. %      See COPYRIGHT.TXT for details.
12. % Modified:
13. %      09 June 2012
14. % Author:
15. %      Per-Olof Persson
16. %      Modifications by John Burkardt
17. % Reference:
18. %
19. %      Per-Olof Persson, Gilbert Strang,
20. %      A Simple Mesh Generator in MATLAB,
21. %      SIAM Review,
22. %      Volume 46, Number 2, June 2004, pages 329-345.
23. %
24. % PARAMETERS:
25. %      INPUTS
26. %      function FD, signed distance function d(x,y).
27. %      function FH, scaled edge length function h(x,y).
28. %      real H0, the initial edge length.
29. %      real BOX(2,2), the bounding box [xmin,ymin; xmax,y-
        max].
30. %      integer ITERATION_MAX, the maximum number of
        iterations.
31. %      The iteration might terminate sooner than this limit, if the
        program
32. %      decides that the mesh has converged.
33. %      real PFIX(NFIX,2), the fixed node positions.
34. % VARARGIN, aditional parameters passed to FD and FH.
35. %
36. %      OUTPUTS
37. %      real P(N,2), the node positions.
38. %      integer T(NT,3), the triangle indices.
39. %
40. % LOCAL PARAMETERS:
```

41. % real GEPS, a tolerance for determining whether a point is "almost"

42. %   inside the region. Setting GEPS = 0 makes this an exact test. The

43. %   program currently sets it to 0.001 * H0, that is, a very small

44. %   multiple of the desired side length of a triangle. GEPS is also

45. %   used to determine whether a triangle falls inside or outside the

46. %   region. In this case, the test is a little tighter. The centroid

47. %   PMID is required to satisfy FD (PMID) <= -GEPS.

48. %

49. dptol = 0.001;

50. ttol = 0.1;

51. Fscale = 1.2;

52. deltat = 0.2;

53. geps = 0.001 * h0;

54. %geps = 0.000000001 * h0

55. deps = sqrt (eps) * h0;

56. iteration = 0;

57. triangulation_count = 0;

58. %

59. % Step 1. Create the initial point distribution by generating a

60. % rectangular mesh in the bounding box.

61. %

62. [x, y] = meshgrid (box(1,1) : h0 : box(2,1), ...

63.   box(1,2) : h0*sqrt(3)/2 : box(2,2));

64. %

65. % Shift the even rows of the mesh to create a "perfect" mesh of

66. % equilateral triangles. Store the X and Y coordinates together as our

67. % first estimate of "P", the mesh points we want.

68. %

69. x(2:2:end,:) = x(2:2:end,:) + h0 / 2;

70. p = [x(:), y(:)];

71. %

72. % Instead of a regular mesh, you can initialize P with random values here.

73. %

74. % Step 2. Remove mesh points that are outside the region, then satisfy the

```
75.%     density constraint. Keep only points inside (or almost inside)
76.% the region, that is, FD(P) < GEPS.
77.%
78.  p = p( feval ( fd, p, varargin{:} ) < geps, : );
79.%
80.% Set r0, the relative probability to keep a point, based on the
     mesh
81.% density function.
82.%
83. r0 = 1 ./ feval ( fh, p, varargin{:} ).^2;
84.%
85.% Apply the rejection method to thin out points according to the
     density.
86.%
87.  p = [ pfix; p(rand(size(p,1),1) < r0 ./ max ( r0 ),: ) ];
88.%
89.% The following obscure commands are written so that:
90.% * We ALWAYS keep the fixed points at the beginning of the
     P array.
91.% * We remove any points which are duplicates of these points.
92.%     That way, we do not later allow the fixed points to move,
93.%     because we know they are at the beginning of the array,
94.%     and DELAUNAYN stops complaining about duplicate
        points.
95.% JVB, 09 June 2012.
96.%
97. [ q, i, j ] = unique ( p, 'rows', 'first' );
98. k = unique ( i );
99. p = p(k,:);
100.
101. N = size ( p, 1 );
102.%
103.% If ITERATION_MAX is 0, we're almost done.
104.%   For just this case, do the triangulation, then exit.
105.%     Setting ITERATION_MAX to 0 means that we can see the
         initial mesh
106.%     before any of the improvements have been made.
107.%
108. if ( iteration_max <= 0 )
109.    t = delaunayn ( p );
```

```
110.    triangulation_count = triangulation_count + 1;
111.    return
112. end
113.
114. pold = inf;
115.
116. while ( iteration < iteration_max )
117.
118.    iteration = iteration + 1;
119.
120.  if ( mod ( iteration, 10 ) == 0 )
121.    fprintf ( 1, ' %d iterations, %d triangulations\n', ...
122.    iteration, triangulation_count );
123.  end
124. %
125. % Step 3. Retriangulation by the Delaunay algorithm.
126. %
127. % Was there large enough movement to retriangulate?
128. %
129. % If so, save the current positions, get the list of
130. % Delaunay triangles, compute the centroids, and keep
131. % the interior triangles (whose centroids are within the region).
132. %
133.    if ( ttol < max ( sqrt ( sum ( ( p - pold ).^2, 2 ) ) / h0 ) )
134.    N = size ( p, 1 );
135.    pold = p;
136.    t = delaunayn ( p );
137.    triangulation_count = triangulation_count + 1;
138.    pmid = ( p(t(:,1),:) + p(t(:,2),:) + p(t(:,3),:) ) / 3;
139.    t = t( feval ( fd, pmid, varargin{:} ) < -geps, : );
140. %
141. % Step 4. Describe each bar by a unique pair of nodes.
142. %
143.    bars = [ t(:,[1,2]); t(:,[1,3]); t(:,[2,3]) ];
144.    bars = unique ( sort ( bars, 2 ), 'rows' );
145. %
146. % Step 5. Graphical output of the current mesh.
147. %
148.    trimesh ( t, p(:,1), p(:,2), zeros(N,1), 'EdgeColor', 'k', 'Line-
        width', 1)
```

```
149.    title ( sprintf ( 'Iteration %d', iteration ) );
150.    view(2), axis equal, axis([-1.25 1.25 -1.25 1.25]), drawnow,
        grid off
151.%
152.% Put a "pause" command here if you'd like to see each new
     mesh.
153.% pause
154.    end
155.%
156.% Step 6. Move mesh points based on bar lengths L and forces F.
157.%
158.% Make a list of the bar vectors and lengths.
159.% Set L0 to the desired lengths, F to the scalar bar forces,
160.% and FVEC to the x, y components of the bar forces.
161.%
162.% At the fixed positions, reset the force to 0.
163.%
164.    barvec = p(bars(:,1),:) - p(bars(:,2),:);
165.    L = sqrt ( sum ( barvec.^2, 2 ) );
166.    hbars = feval ( fh, (p(bars(:,1),:)+p(bars(:,2),:))/2, varargin{:} );
167.    L0 = hbars * Fscale * sqrt ( sum(L.^2) / sum(hbars.^2) );
168.    F = max ( L0 - L, 0 );
169.    Fvec = F ./ L * [1,1] .* barvec;
170.    Ftot=full(sparse(bars(:,[1,1,2,2]),ones(size(F))*[1,2,1,2],[Fve
        c,-Fvec],N,2));
171.    Ftot(1:size(pfix,1),:) = 0;
172.    p = p + deltat * Ftot;
173.%
174.% Step 7. Bring outside points back to the boundary.
175.%
176.% Use the numerical gradient of FD to project points back to the
     boundary.
177.%
178.d = feval ( fd, p, varargin{:} );
179.ix = d > 0;
180.dgradx = ( feval(fd,[p(ix,1)+deps,p(ix,2)],varargin{:}) - d(ix) ) / deps;
181.dgrady = ( feval(fd,[p(ix,1),p(ix,2)+deps],varargin{:}) - d(ix) ) / deps;
182.p(ix,:) = p(ix,:) - [ d(ix) .* dgradx, d(ix) .* dgrady ];
183.%
184.% Step 8. Termination criterion: All interior nodes move less
     than dptol
```

```
185. %              (scaled).
186. %
187.    if (max(sqrt( sum ( deltat * Ftot ( d < -geps,:).^2, 2 ) ) / h0 )<
        dptol)
188.      break;
189.    end
190.  end
191.  return
192. end
```

APPENDIX 5C MATLAB CODE FOR THE M-FILES USED IN DISTMESH_2D

5C.1   INTRODUCTION

This appendix presents the m-files for the mesh generator distmesh_2d that are called by the six different meshes that are formed by the main driver program.

5C.2   MATLAB CODE

The separate m-files are listed here. They are also contained in Persson and Strang (2004) in the main text with the exception of dparabola.m.

5C.2a   ddiff.m

```
1 function d= ddiff(d1,d2) % Difference
2 d=max(d1,-d2);
```

5C.2b   dcircle.m

```
1 function d=dcircle(p,xc,yc,r) % Circle
2 d=sqrt((p(:,1)-xc).^2+(p(:,2)-yc).^2)-r;
```

5C.2c dintersection.m

```
1 function d= dintersection(d1,d2) % Intersection
2 d=max(d1,d2);
```

5C.2d   dparabola.m

```
1 function d=dparabola(p,xc,yc)
2 aa=0.3; % Minor axis.
3 bb=0.5; % Major axis.
4 d=sqrt((((p(:,1)-xc)/aa).^2+((p(:,2)-yc)/bb).^2)-1;
```

5C.2e   drectangle.m

```
1 function d=drectangle(p,x1,x2,y1,y2) % Rectangle
2 d=-min(min(min(-y1+p(:,2),y2-p(:,2)), ...
3 -x1+p(:,1)),x2-p(:,1));
```

5C.2f   dunion.m

```
1 function d=dunion(d1,d2)
2 d=min(d1,d2);
```

5C.2g   huniform.m

```
1 function h=huniform(p,varargin) % Uniform h(x,y) disstribution
2 h=ones(size(p,1),1);
```

CHAPTER 6

FORMATION OF A FINITE ELEMENT MODEL OF THE KIRSCH PROBLEM

6.1   INTRODUCTION

This chapter provides a compact overview of the finite element method and the nature of its results. This is accomplished by creating and solving the Kirsch problem with a sequence of finite element models that are formed with different sized elements. Because of its simplicity and the severity of its stress concentrations, this problem makes an ideal platform for the developments presented here.

Figure 6.1 shows a finite element model of this problem formed with a coarse mesh of three-node elements. This model consists of a square panel with a centrally located circular hole. It is loaded in tension at both ends with identical uniformly distributed loads. The loads are indicated by the triangles on the right- and left-hand boundaries in Figure 6.1. Stress concentrations exist at both the top and bottom of the circular cutout.

The stress distributions produced by the applied loads are compared for the sequence of problems in order to identify salient characteristics of finite element solutions. The two most significant characteristics are: (1) jumps or discontinuities that exist in the strains between adjacent elements when the finite element result does not exactly represent the actual solution and (2) the magnitude of the jumps correlate with the level of error in the approximate finite element solution. These characteristics provide insights that lead to the creation of the error estimators and refinement guides that are developed in the chapters that follow.

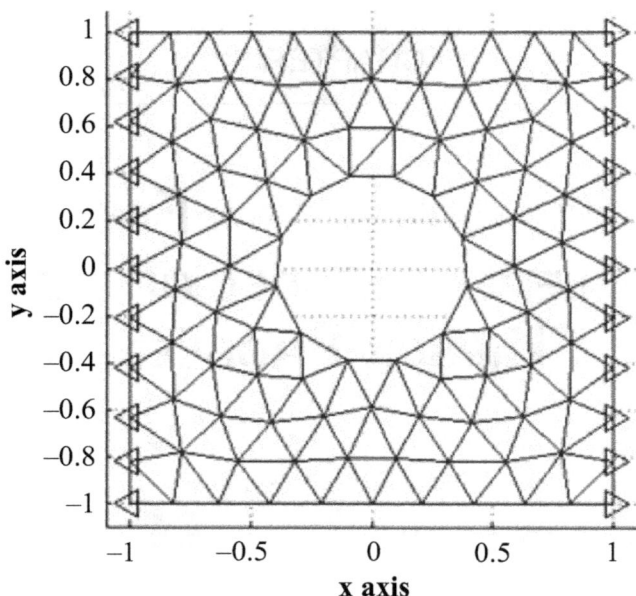

Figure 6.1. Initial finite element model.

6.2 AN OUTLINE OF THE FINITE ELEMENT MODEL FORMULATION PROCESS

The finite element model of the Kirsch problem is formed and solved by integrating and extending the contents of the previous chapters. The three-node elements used to form the model are developed using the procedures based on the physically interpretable notation that is developed in Chapter 3 and applied in Chapter 4. The mesh generation program presented in Chapter 5 is used to create the meshes for the problems solved in this chapter. The individual stiffness matrices in the model are based on the nodal locations produced by the mesh generator. These elements are assembled to form the unrestrained global stiffness matrix using the procedures presented in Chapter 2.

Finally, the unrestrained model is completed by applying the uniformly distributed loads to the two ends of the panel. At this stage of the model formulation, the finite element representation is the following:

$$[K] \{d\} = \{F\} \qquad (6.1)$$

where $[K]$ is the global, unrestrained stiffness matrix, $\{d\}$ is the vector of nodal displacements, and $\{F\}$ is the vector of applied loads.

In physical terms, an unrestrained stiffness matrix for a two-dimensional problem can experience three rigid body motions, displacements in the x and y directions, and a rotation around the z axis. In mathematical terms, this means that the unrestrained stiffness matrix cannot be inverted so the problem cannot be solved for the displacements.

Since the unknowns in Equation 6.1 are the nodal displacements, the stiffness matrix must be inverted so that the displacements can be found for this statics problem. In order to make this possible, the finite element model must be constrained so that it cannot experience any of the three rigid body motions. The stiffness matrix is constrained by forcing three of the nodal displacements in the unrestrained finite element model to have a displacement equal to zero.

The constraints are applied to degrees of freedom that naturally have zero displacements because of the symmetries in the geometry and loading of the problem. Since the displacements that are forced to have zero displacement would have zero displacements anyway, the deformations and, hence, the stress concentrations are not affected by these constraints.

After the stiffness matrix is restrained and inverted, the final finite element model has the following form:

$$\{d\} = [K]^{-1} \{F\} \tag{6.2}$$

where [K] is the global, restrained stiffness matrix, {d} is the vector of nodal displacements, and {F} is the vector of applied loads.

Once the displacements are found, the elemental strains are extracted. This is accomplished by using the strain–displacement relation that was used in the formulation of the finite element using Equations 4.5 and 4.13. Then the strains are inserted into the stress–strain relationship to produce the elemental stresses.

The three stress components for the elements on the boundary of the interior circle are shown in Figure 6.2. The stresses are presented with respect to a coordinate system that is rotated so that it is parallel and perpendicular to the edge of each element. As can be seen in Figure 6.2a, two stress concentrations exist on the boundary of the internal hole in the normal stress component that is tangential to the edge of the element.

The primary characteristic of a finite element solution that is critical to the development of error estimators and mesh refinement guides can be seen in Figure 6.2. Specifically, there are jumps in all three stress components between the adjacent elements. These discontinuities exist because the strain distributions that an individual element is attempting to capture

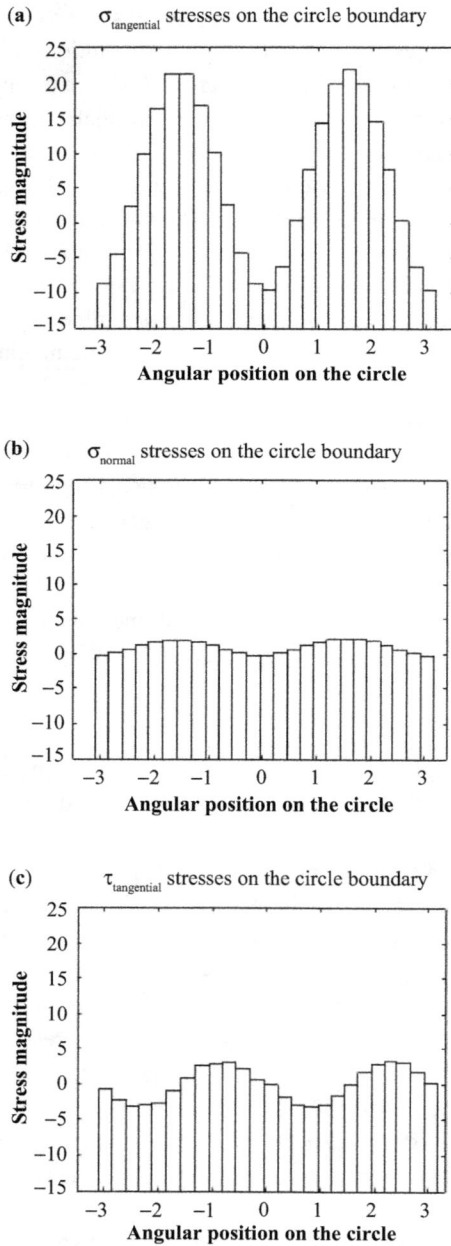

Figure 6.2. Stress components on the boundary of the circular hole: (a) tangential normal stress, (b) perpendicular normal stress, and (c) tangential shear stress.

are more complex than a single element is capable of representing. In mathematical terms, the Taylor series representation of the actual stress distribution is of a higher order than that which a three-node constant strain element is capable of representing.

Note that the interelement jumps are larger for the critical stress shown in Figure 6.2a than for the other two stress components. This is the case because the gradients or the rates of change in the actual strain are higher for this component than for the other two components. Although the interelement jumps for the stress components shown in Figures 6.2b and 6.2c are smaller, they also represent modeling errors. Since these components represent the stress on an unloaded boundary, they should be equal to zero. However, as can be seen in Figures 6.2b and 6.2c, these stresses are *not equal* to zero.

In order to demonstrate how a finite element results behaves as the model is improved, the stresses for a sequence of models will be compared later in this chapter. Specifically, we will see that the interelement jumps in the stresses decrease and the stress representations improve as elements are added in regions with high strain gradients. This characteristic provides the basis for the error estimators and refinement guides developed in the next two chapters.

As previously noted, the interelement jumps and the associated errors derive from the inability of an element to represent the higher-order strain gradients that exist in the actual solution. It is this direct relationship between the source of the modeling errors and the strain gradient notation that gives the physically based notation presented here its usefulness and power. Consequently, this notation will be used to define the metrics that quantify the errors in a finite element solution as well as the level of refinement in the mesh needed to produce acceptable results.

6.3   FORMULATION OF A THREE-NODE CONSTANT STRAIN ELEMENT

This section applies the seven-step procedure presented in Chapter 4 to form the stiffness matrix for a three-node finite element. This formulation is based on the strain gradient notation that was introduced in Chapter 3.

STEP 1—IDENTIFICATION OF THE DISPLACEMENT INTERPOLATION POLYNOMIALS

Example configurations of three-node elements are shown in Figure 6.3. The standard counter-clockwise numbering of the element nodes is shown

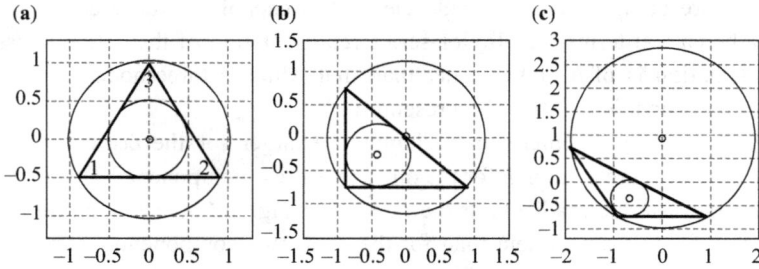

Figure 6.3. Sample configurations of three-node elements: (a) equilateral triangle cond. no. = 0.5358, (b) right triangle cond. no. = 0.2636, and (c) oblique triangle cond. no. = 0.0853.

in Figure 6.3a. The displacement interpolation polynomials expressed in strain gradient notation for a three-node element are the following:

$$u(x, y) = (u_{rb})_0 + (\varepsilon_x)_0 x + (\gamma_{xy} / 2 - r_{rb})_0 y$$
$$v(x, y) = (v_{rb})_0 + (\gamma_{xy} / 2 + r_{rb})_0 x + (\varepsilon_y)_0 y \tag{6.3}$$

The displacement interpolation polynomials and the modeling capabilities for a three-node element are a subset of the interpolation polynomials for the six-node element that were identified in Equations 4.1 and 4.2. The displacement polynomials for a six-node element are complete quadratic representations. As can be seen in Equation 6.3, the displacement representations for a three-node element are complete linear polynomials. They do not contain the quadratic displacement representations. This means that a three-node element can only represent linear changes in the displacements. As we will see in the next step, this means that a three-node element can only represent constant strains.

The following six linearly independent strain states contained in Equation 6.3 are capable of being represented by a three-node element:

$$(u_{rb})_0 \quad (v_{rb})_0 \quad (r_{rb})_0 \quad (\varepsilon_x)_0 \quad (\varepsilon_y)_0 \quad (\lambda_{xy})_0 \tag{6.4}$$

These terms indicate that a three-node element can represent the three rigid body motions and constant strain values for the three strain components. The displacements in an element are produced by a linear combination of these six strain states.

It must be noted that the ability to represent the three rigid body motions and the three constant strain states satisfies the convergence criteria for a two-dimensional finite element. This means that with enough refinement any problem can be represented exactly with any planar element that satisfies the convergence criteria. In physical terms, this means that, if required because of the complexity of the problem, very physical point in a problem can be represented by a single element. Since a single point can only experience rigid body motions and constant strains, these conditions are within the capabilities of a three-node element.

STEP 2—STRAIN FORMULATION AND EVALUATION

The strain modeling capabilities and the limitations of these capabilities for a three-node element are presented in this step. The strain representations for a three-node element are identified by applying the linear elasticity definitions of the strain components to the displacement interpolation polynomials given by Equation 6.3 to produce the following:

$$\varepsilon_x\,(x,y) = \partial u/\partial x = (\varepsilon_x)_0$$
$$\varepsilon_y\,(x,y) = \partial v/\partial y = (\varepsilon_y)_0 \qquad\qquad (6.5)$$
$$\gamma_{xy}\,(x,y) = \partial v/\partial x + \partial u/\partial y = (\gamma_{xy})_0$$

Equation 6.5 shows that a three-node element can only represent constant strains. This means that an individual three-node element cannot accurately represent an actual strain distribution that is more complex. For example, a three-node element cannot represent a linear change in the ε_x strain component because the strain model for $\varepsilon_x(x,y)$ does not include the $\varepsilon_{x,x}x$ term in its representation. In contrast, a six-node linear strain element can represent linear strain distributions (see Equation 4.3). However, a six-node element has its limitations: it cannot represent quadratic strain distributions because its displacement representations do not contain cubic terms.

Since a three-node element can only represent constant strain distributions, a finite element model formed with these elements cannot represent more complex strain distributions exactly. A finite element model formed with three-node elements represents the changes or gradients in the actual solution with changing values of constant strains in the adjacent elements, as shown in Figure 6.2. It is this attempt to capture the exact result with a finite number of strain states that produces the interelement jumps in the finite element stress and strain representations.

When the strain representations for a three-node element are cast in the matrix form that is used to express the strain energy expression in the formulation of the element stiffness matrix, the result is the following:

$$\{\varepsilon\} = [T]\{\varepsilon,\} \tag{6.6}$$

where $\{\varepsilon\}^T = [\ \varepsilon_x \ \varepsilon_y \ \gamma_{xy}\ ]$

$\{\varepsilon,\}^T = [(u_{rb})_0 \ (v_{rb})_0 \ (r_{rb})_0 \ (\varepsilon_x)_0 \ (\varepsilon_y)_0 \ (\gamma_{xy})_0]$

$[T] = [\,[T_0]\ [T_\varepsilon]\,]$ and

$$[T_0] = \begin{bmatrix} 0 & 0 & 0 \\ 0 & 0 & 0 \\ 0 & 0 & 0 \end{bmatrix} \quad \text{and} \quad [T_\varepsilon] = \begin{bmatrix} 1 & 0 & 0 \\ 0 & 1 & 0 \\ 0 & 0 & 1 \end{bmatrix}$$

In this representation of the strain models, the strain expressions given by Equation 6.5 have been augmented with the rigid body motions of the element. The rigid body terms are needed in the strain representation so that the effects of rigid body motion are included in the stiffness matrix for the element.

By their very name, the rigid body motions do not deform the element so no strain energy is added when an element experiences a rigid body motion. The fact that the rigid body motions do not influence the strains is seen by their introduction into Equation 6.6 with the null matrix designated as $[T_0]$.

STEP 3—FORMULATION OF THE STRAIN ENERGY EXPRESSION

The discrete form of the strain energy for a two-dimensional domain is formed by substituting the coordinate transformation given by Equation 6.6 into the continuous expression of the strain energy as follows:

$$\begin{aligned} SE &= \tfrac{1}{2}\int_\Omega \{\varepsilon\}^T [E]\{\varepsilon\}\, d\Omega \\ &= \tfrac{1}{2}\int_\Omega \{\varepsilon,\}^T [T]^T [E][T]\{\varepsilon,\}\, d\Omega \\ &= \tfrac{1}{2}\{\varepsilon,\}^T \left[\int_\Omega [T]^T [E][T]\, d\Omega\right]\{\varepsilon,\} \\ &= \tfrac{1}{2}\{\varepsilon,\}^T \bar{U}\{\varepsilon,\} \end{aligned} \tag{6.7}$$

where

$$[E] = \frac{E}{\left(1-v^2\right)}\begin{bmatrix} 1 & \frac{1}{2} & 0 \\ v & 1 & 0 \\ 0 & 0 & \alpha \end{bmatrix}$$

$$\alpha = \frac{1-v}{2}$$

E = Young's Modulus

v = Poisson's ratio

When the expression for $\bar{U}$ is expanded using the partitioned transformation matrix of Equation 6.6, the result is the following:

$$\bar{U} = \int_\Omega [T]^T [E][T] d\Omega$$

$$= \int_\Omega \begin{bmatrix} [T_0]^T [E][T_0] & [T_0]^T [E][T_\varepsilon] \\ [T_\varepsilon]^T [E][T_0] & [T_\varepsilon]^T [E][T_\varepsilon] \end{bmatrix} d\Omega$$

$$= \int_\Omega \begin{bmatrix} [0] & [0] \\ [0] & \bar{U}_{22} \end{bmatrix} d\Omega \qquad (6.8)$$

Since the matrix $[T_0]$ is the null matrix, the partitions of Equation 6.8 containing this matrix are equal to zero when they are multiplied out. As a result, the only nonzero partition in this strain energy expression is $\bar{U}_{22}$.

The creation and recognition of these partitions in the strain energy expression containing only zeros are only possible because the rigid body motions are explicitly identified in strain gradient notation. As we will see in the next step, this simplifies the element formulation process because the number of integrals that must be evaluated is significantly reduced.

STEP 4—INTEGRATION OF THE STRAIN ENERGY EXPRESSION

When the nonzero submatrix $\bar{U}_{22}$ of Equation 6.8 is expanded, we have the following:

$$\bar{U}_{22} = \int_\Omega [T_\varepsilon]^T [E][T_\varepsilon] d\Omega$$

$$= \frac{tE}{(1-v^2)}\begin{bmatrix} I_1 & v I_1 & 0 \\ v I_1 & I_1 & 0 \\ 0 & 0 & \alpha I_1 \end{bmatrix} \qquad (6.9)$$

where

$$t = \text{thickness}$$

$I_1 = \int_\Omega d\Omega$ which represents the area of the triangle

As can be seen in Equation 6.9, every entry contains the term I_1 which is the area of the finite element. This contrasts to higher-order finite elements that require the evaluation of other integrals in addition to the area integral.

The area of the element can be determined symbolically as a function of the three nodal points that define the triangular element with a determinant as follows:

$$I_1 = \frac{1}{2} \begin{Vmatrix} 1 & 1 & 1 \\ x_1 & x_2 & x_3 \\ y_1 & y_2 & y_3 \end{Vmatrix}$$

$$= (x_2\, y_3 + x_3\, y_1 + x_1\, y_2 - x_2\, y_1 - x_1\, y_3 - x_3\, y_2)/2 \qquad (6.10)$$

This determinant finds the volume of an area bounded by two sets of parallel lines (a rectangular parallelepiped) with a thickness of one unit that consists of two triangles that are identical to the shape of the finite element. When the determinant is divided by two as is done in Equation 6.10, the result is the area of the triangular element.

STEP 5—FORMULATION OF THE TRANSFORMATION TO NODAL COORDINATES

As can be seen in Equation 6.7, the strain energy expression is given as a function of strain gradient coordinates. In order to be able to assemble the individual stiffness matrices into a global model, the individual stiffness matrices had to be expressed in terms of Cartesian coordinates in the global x-y system. The need for this coordinate transformation was demonstrated in Chapter 2 when the stiffness matrix for the hexagonal truss was assembled from the truss elements.

Thus, we must transform the strain energy expression given by Equation 6.7 from strain gradient coordinates to displacement coordinates. The transformation required for this change from physically interpretable to global Cartesian coordinates is formed from Equation 6.3.

When Equation 6.3 is cast in matrix form, we have the following:

$$\begin{Bmatrix} u_i \\ v_i \end{Bmatrix} = \begin{bmatrix} 1 & 0 & -y_i & x_i & 0 & y_i/2 \\ 0 & 1 & x_i & 0 & y_i & x_i/2 \end{bmatrix} \{\varepsilon,\} \tag{6.11}$$

where $\{\varepsilon,\}$ is defined in Equation 6.6.

The transformation for expressing the strain energy given in the physically interpretable coordinates by Equation 6.7 to nodal coordinates is formed by evaluating Equation 6.11 at each of the three nodal points of the element to give the following:

$$\{d\} = [\Phi]\{\varepsilon,\} \tag{6.12}$$

where

$$\{d\} = \begin{bmatrix} u_1 & u_2 & u_3 & v_1 & v_2 & v_3 \end{bmatrix}^T \text{ and}$$

$$[\Phi] = \begin{bmatrix} 1 & 0 & -y_1 & x_1 & 0 & y_1/2 \\ 1 & 0 & -y_2 & x_2 & 0 & y_2/2 \\ 1 & 0 & -y_3 & x_3 & 0 & y_3/2 \\ 0 & 1 & x_1 & 0 & y_1 & x_1/2 \\ 0 & 1 & x_2 & 0 & y_2 & x_2/2 \\ 0 & 1 & x_3 & 0 & y_3 & x_3/2 \end{bmatrix}$$

Note that the nodal displacements in the x direction precede the displacements in the y direction. In this form, the structure of the rigid body displacements is clear. The first column represents the motion of the three nodes in the element when it is experiencing a rigid body displacement in the x direction with a magnitude of u_{rb}. As can be seen by the locations of the ones and zeros, the nodes will move in the x direction, not the y direction.

The normal strains are seen to depend on only the x and y displacements. Finally, rigid body rotation in the third column and the shear strain representation in the sixth column can be seen to depend on the displacements in both the x and y directions. Examples of these rigid body motions and deformations for the hexagonal truss are presented in Figure 3.5.

The inverse of the transformation from nodal displacements to strain gradient quantities is used in the element formulation process.

This is seen in Equation 6.7. When Equation 6.12 is inverted, we have the following transformation from strain gradient quantities to nodal displacements:

$$\{\varepsilon,\} = [\Phi]^{-1}\{d\} \tag{6.13}$$

where

$$[\Phi]^{-1} = \begin{bmatrix} \Phi_{11}^{-1} & \Phi_{12}^{-1} \\ \Phi_{21}^{-1} & \Phi_{22}^{-1} \end{bmatrix}$$

The notation in Equation 6.13 is somewhat ambiguous. However, the partitions of this matrix are needed later to simplify the element formulation process. In order to remove this ambiguity, the meaning of these partitions will be clarified. The key point is to note, for example, that $\Phi_{11}^{-1} \neq [\Phi_{11}]^{-1}$. The matrix Φ_{11}^{-1} is extracted from $[\Phi]^{-1}$ after $[\Phi]$ is inverted. Similarly, when the transpose of this partition is formed, it is equal to $[\Phi_{11}^{-1}]^{T}$. It is not equal to $\Phi_{11}^{-T} \neq [\Phi_{11}]^{-T}$.

This partition of $[\Phi]$ is isolated because it is associated with the rigid body motions. Since the rigid body motions do not produce any strain energy, the terms containing this partition will be zero in the strain energy expression. Since the terms containing this partition are known to be zero, they do not need to be explicitly computed. As a result, the computation required to form the stiffness matrix for an element is reduced.

STEP 6—TRANSFORMATION OF THE STRAIN ENERGY TO NODAL DISPLACEMENTS

The strain energy expression given by Equation 6.7 will now be transformed to nodal displacement coordinates by substituting the coordinate transformation given by Equation 6.13. When this is done, we have the following:

$$\begin{aligned} SE &= \tfrac{1}{2}\{\varepsilon,\}^{T} \bar{U}\{\varepsilon,\} \\ &= \tfrac{1}{2}\{d\}^{T}[\Phi]^{-T} \bar{U}[\Phi]^{-1}\{d\} \\ &= \tfrac{1}{2}\{d\}^{T}[K]\{d\} \end{aligned} \tag{6.14}$$

As noted in Chapter 4, the matrix $\bar{U}$ has been evaluated in step 4 and $[K]$ is symmetric.

STEP 7—APPLICATION OF THE PRINCIPLE OF MINIMUM POTENTIAL ENERGY

In Chapter 4, the principle of minimum potential energy was used to identify the theoretical form of the element stiffness matrix. In this version of step 7, we will evaluate the stiffness matrix for a three-node element in symbolic form using the theoretical result presented in Chapter 4. This will be accomplished by expanding the definition of [K] given in Equation 6.14 as follows:

$$[K] = [\Phi]^{-T} \bar{U} [\Phi]^{-1}$$

$$= \begin{bmatrix} \Phi_{11} & \Phi_{12} \\ \Phi_{21} & \Phi_{22} \end{bmatrix}^{-T} \begin{bmatrix} \bar{U}_{11} & \bar{U}_{12} \\ \bar{U}_{21} & \bar{U}_{22} \end{bmatrix} \begin{bmatrix} \Phi_{11} & \Phi_{12} \\ \Phi_{21} & \Phi_{22} \end{bmatrix}^{-1}$$

$$= \begin{bmatrix} \Phi_{11}^{-T} & \Phi_{12}^{-T} \\ \Phi_{21}^{-T} & \Phi_{22}^{-T} \end{bmatrix} \begin{bmatrix} [0] & [0] \\ [0] & \bar{U}_{22} \end{bmatrix} \begin{bmatrix} \Phi_{11}^{-1} & \Phi_{12}^{-1} \\ \Phi_{21}^{-1} & \Phi_{22}^{-1} \end{bmatrix}$$

$$= \begin{bmatrix} [0] & \Phi_{12}^{-T} \bar{U}_{22} \\ [0] & \Phi_{22}^{-T} \bar{U}_{22} \end{bmatrix} \begin{bmatrix} \Phi_{11}^{-1} & \Phi_{12}^{-1} \\ \Phi_{21}^{-1} & \Phi_{22}^{-1} \end{bmatrix}$$

$$= \begin{bmatrix} \Phi_{12}^{-T} \bar{U}_{22} \Phi_{21}^{-1} & \Phi_{12}^{-T} \bar{U}_{22} \Phi_{22}^{-1} \\ \Phi_{22}^{-T} \bar{U}_{22} \Phi_{21}^{-1} & \Phi_{22}^{-T} \bar{U}_{22} \Phi_{22}^{-1} \end{bmatrix} \tag{6.15}$$

When the final expression for the stiffness matrix is extracted from Equation 6.15, we have the following:

$$[K] = \begin{bmatrix} \Phi_{12}^{-T} \bar{U}_{22} \Phi_{21}^{-1} & \Phi_{12}^{-T} \bar{U}_{22} \Phi_{22}^{-1} \\ \Phi_{22}^{-T} \bar{U}_{22} \Phi_{21}^{-1} & \Phi_{22}^{-T} \bar{U}_{22} \Phi_{22}^{-1} \end{bmatrix} \tag{6.16}$$

The formulation of the three-node element stiffness matrix with symbolic computation is presented in detail in Appendix 6A. This appendix has two functions. The first is to clarify the strain gradient stiffness matrix formulation with a concrete example. The second is to provide an example of symbolic computation for those unfamiliar with this powerful tool (Dow 1999, 2012).

6.4   ELEMENT CONDITION NUMBERS

Before continuing the formulation of the finite element model, let us consider the *quality* of a finite element stiffness matrix. In the previous chapter, the quality of a triangle was measured in terms of its geometry. The quality was measured by the ratio of the diameter of the inscribed circle to the diameter of the circumscribed circle for an element. This measure essentially quantifies how close a triangle is to a circle. With this metric, an equilateral triangle has the largest condition number, so the closer a triangle is to an equilateral triangle the better it is considered to be.

The quality of the geometry of a triangle was of interest in the previous chapter because one of the goals of the mesh generator is to produce a mesh with triangles that are close to being equilateral triangles. This goal is significant for the finite element method because the closer a triangle is to an equilateral triangle, the better are its numerical characteristics.

In this section, the quality of a finite element stiffness matrix is measured by the ratio of the minimum nonzero eigenvalue to that of the maximum eigenvalue. This measure, called the *condition number*, relates to the difference between the largest and the smallest number in the stiffness matrix. Depending on the number of significant figures in the calculations, the condition number is related to the accuracy of the displacements found from Equation 6.1 due to the round-off in the computations. The two measures of quality are compared in Table 6.1.

As can be seen, the triangles shown in Figure 6.3 are ranked in the same order by both metrics. However, the condition number for the stiffness matrices deteriorates more rapidly as the configuration departs from being an equilateral triangle.

At this point, a comment concerning the need to constrain the stiffness matrix in Equation 6.1 can be made. The stiffness matrix for a single planar finite element is capable of representing three rigid body motions. In a similar manner, the unrestrained stiffness matrix for a finite element model is capable of representing three rigid body motions.

The existence of the three rigid body motions can be seen when an unrestrained stiffness matrix of an element or a full model is introduced into an eigenvalue solver. For the case of unrestrained stiffness matrices, three of the eigenvalues are found to be equal to zero. When the physical system is given the displacements associated with the zero eigenvalues, it is not deformed. In other words, the system is experiencing *rigid body motion*.

Table 6.1. Triangle quality metrics

Triangle type	Condition number	Geometric quality
Equilateral	0.5358	1.0
Right	0.2636	0.8168
Oblique	0.0853	0.4173

One characteristic of the stiffness matrices for planar elements is worth noting. The stiffness matrices for planar elements with different sizes, but the same configuration, are identical. For example, the stiffness matrix for an equilateral triangle is the same regardless of its size. In other words, the stiffness matrices for planar elements are *scale-free* (Dow and Byrd 1981). The demonstration of this counterintuitive characteristic is left as an exercise at the end of this chapter.

Since the stiffness matrices for planar elements are scale-free, one can conceive of the usefulness of finite element stress concentration libraries. This idea which was suggested in Dow and Byrd (1981) can be easily implemented with a mesh generator and the error estimators and the mesh refinement guides being developed here. With such a library, highly accurate models of stress concentrations with a reduced number of degrees of freedom can be inserted in larger problems.

6.5 CREATION AND APPLICATION OF THE FINITE ELEMENT MODEL

In this section, the MATLAB program that creates the finite element models used in this presentation is outlined. All of the developments presented here are demonstrated with the Kirsch problem shown in Figure 5.2. The presentation consists of an outline of the seven steps that comprise the program. The outline, in conjunction with the mesh generation program presented in Chapter 5 and the element generation program presented in Appendix 6A, is presented in order to assist the reader if they choose to create a similar program.

The presentation is illustrated with a model formed from a mesh with a nodal spacing specified as $h_0 = 0.2$ in the mesh generation program.

The program consists of the following seven steps:

1. Mesh generation
2. Identification of the nodal displacement constraints
3. Formulation of the unrestrained stiffness matrix
4. Formulation of the restrained stiffness matrix
5. Identification and introduction of applied loads
6. Solution of the model for the nodal displacements
7. Extraction of elemental and nodal strains

These seven steps will be discussed in turn.

STEP 1—MESH GENERATION

The procedure for forming a mesh was presented in detail in Chapter 5. As was discussed at length, the output of the mesh generator consists of the locations of the nodal points that make up the model and the element topology. The x and y coordinates of the nodal points are contained in the rows of the matrix identified as p in the mesh generator. The element topology is given in the rows of the matrix identified as t and consists of the three nodal numbers that identify the individual triangles.

The results of the mesh generator are presented visually in Figure 6.4 with plots of the mesh with the node and element numbers. As can be seen in Figure 6.4a, the four corners of the mesh, which are defined as the first four fixed points in the mesh generator, are identified as nodes 1 to 4. The coordinates for these four nodal locations are $x = \pm 1$ and $y = \pm 1$.

The element topology can be extracted from Figure 6.4. For example, the topology of element 60 located in the lower left-hand corner is given by the identification number of its three nodes, namely, $t(60, 3) = 1, 9, 20$.

STEP 2—IDENTIFYING NODAL DISPLACEMENT CONSTRAINTS

In order to invert the stiffness matrix for this statics problem, it must be constrained so that the three rigid body motions cannot occur. Furthermore, in order to eliminate any effect of the constraints on the stress concentration being studied, the constraints must be chosen so they do

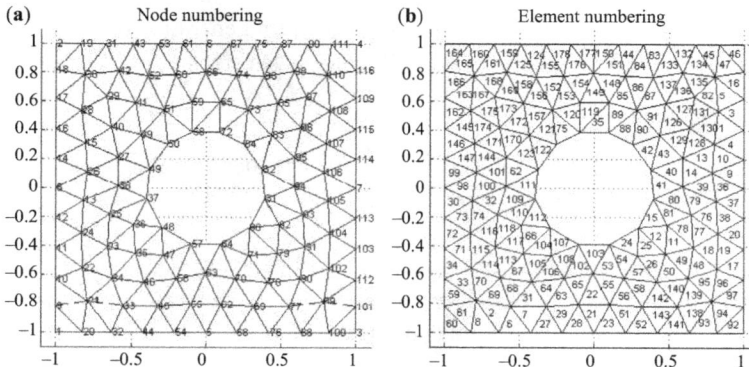

Figure 6.4. Graphical output of the mesh generation program: (a) mesh with node numbering and (b) mesh with element numbering.

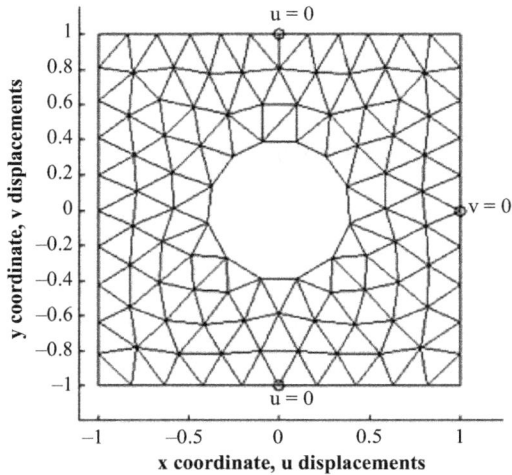

Figure 6.5. Mesh with nodal constraints.

not affect the stresses or strains in the problem. This is accomplished by constraining three coordinates that have no motion because of the configuration and loading of the problem. These constraints are shown in Figure 6.5.

In this case, the following three constraints are chosen. The displacements in the x direction of nodes 5 and 6 that are centered on the top and bottom of the square panel are the first two constraints. These constraints are shown in Figure 6.5 with the nodes indicated by the circles and denoted by $u = 0$. The displacement of these nodal coordinates will be zero as a result of the symmetry of the geometry and the loading of the problem.

Similarly, the third constraint is chosen as the displacement of a node that will have zero displacement in the y direction. This constraint consists of the displacement in the y direction of node 7 that is designated by the circle on the right-hand side of the panel in the center at $x = + 1.0$ and $y = 0.0$. The displacement is designated as $v = 0$. These three constraints will not allow the model to exhibit either rigid body displacements in the two coordinate directions or rotation around the z axis.

STEP 3—FORMULATION OF THE UNRESTRAINED STIFFNESS MATRIX

The assembly of the unrestrained global stiffness matrix for a truss, one element at a time, was developed, described, and demonstrated in

Chapter 2. The procedure for creating the stiffness matrix for two- and three-dimensional finite element models is similar to that used to form truss models. In fact, the formulation process for finite element models is even simpler.

The feature of a multidimensional finite element that makes the assembly of a global stiffness matrix simpler than for a truss model can be seen in Figures 6.4a or 6.4b. This figure shows that the orientation of the individual elements is defined in the mesh generation process. In other words, the elemental stiffness matrices for these two-dimensional elements are computed directly in terms of global coordinates. This contrasts to the stiffness matrices for truss elements that are initially computed in local coordinates. Then, they are rotated so that they are expressed in global coordinates.

STEP 4—FORMULATION OF THE RESTRAINED STIFFNESS MATRIX

As discussed earlier, the unrestrained stiffness matrix must be constrained so it can be inverted. As we saw in Equation 6.2 and will see in step 6, the stiffness matrix must be inverted if we are to solve for the displacements. In step 2, we identified the nodal displacements that will be restrained in order to make the stiffness matrix invertible.

The restraining of the three coordinates simply means that we have defined the displacements of these degrees of freedom to be zero. In other words, we do not need to find these three displacements because they have been specified as zero. It is the incorporation of these known displacements that provides the basis for forming the restrained stiffness matrix.

For convenience in the following development, let us assume that the three constrained coordinates are the last three coordinates in the unconstrained stiffness matrix. This assumption allows us to partition the unrestrained stiffness matrix in a compact configuration because the final three displacements are equal to zero. Since the final three displacements are equal to zero, we can partition the unrestrained stiffness matrix as follows:

$$\begin{bmatrix} K_{11} & K_{12} \\ K_{21} & K_{22} \end{bmatrix} \begin{Bmatrix} d_1 \\ 0 \end{Bmatrix} = \begin{Bmatrix} F_1 \\ F_2 \end{Bmatrix} \tag{6.17}$$

In this expression, the bottom partition contains the three rows associated with the three constrained degrees of freedom. Similarly, the last three columns are associated with the three constrained coordinates.

Thus, if the unrestrained stiffness matrix has n degrees of freedom, the four partitions in Equation 6.17 will have the following sizes: matrix K_{11} is a $(n-3) \times (n-3)$ matrix, matrix K_{22} is a 3×3 matrix, matrix K_{12} is a $3 \times (n-3)$ matrix, and matrix K_{21} is a $(n-3) \times 3$ matrix.

As a result of this partitioning, the top partition of Equation 6.17 can be rearranged as follows:

$$\left[K_{11}\right]\{d_1\} + \left[K_{12}\right]\{0\} = \{F_1\}$$
$$\left[K_{11}\right]\{d_1\} = \{F_1\} - \left[K_{12}\right]\{0\}$$
$$\left[K_{11}\right]\{d_1\} = \{F_1\} \qquad (6.18)$$

Before continuing, let us interpret Equation 6.18. The partition $[K_{11}]$ of the unrestrained stiffness is the restrained stiffness matrix. It cannot undergo any rigid body motion, so all of its eigenvalues are nonzero. In other words, the partition $[K_{11}]$ is not singular, so it can be inverted. This means that we can now solve for the displacements d_1 after we identify the load vector F_1.

Note that if the constrained displacements are not zero and we denote them as d_2, they would enter the solution as *fictitious* applied loads that are equal to $-[K_{12}]\{d_2\}$. Furthermore, if we desire, we can find the reactions at the location of the constrained displacements. The reaction forces are available from the lower partition of Equation 6.17 as follows:

$$\{F_2\} = \left[K_{21}\right]\{d_2\} \qquad (6.19)$$

Before we can solve for the displacements in the Kirsch problem, we must first define the applied loads for this problem.

STEP 5—IDENTIFICATION AND INTRODUCTION OF THE APPLIED LOADS

The loading of the Kirsch problem consists of equal and opposite distributed loads at the two ends of the panel. This loading is shown schematically in Figure 6.6 as triangles on the end nodes. The triangles on the right-hand end, which point to the right indicate tensile loads. In order to put the panel in equilibrium, a distributed load of equal magnitude is applied in the negative x direction. This is indicated on the figure by the triangles on the left-hand end, which point in the negative x direction.

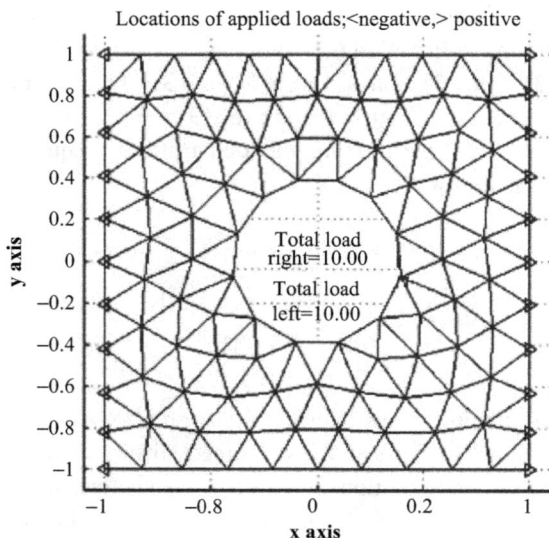

Figure 6.6. Mesh with applied loads identified.

In the problems solved here, the total distributed load consists of 10 units at each end of the panel. The total loads on the two ends of the panel are indicated in the center of Figure 6.6.

In a three-node finite element, a uniformly distributed load along an edge is evenly divided between the two nodes. This means that the loads on the top and bottom nodes at $x = \pm 1$ and $y = \pm 1$ are equal to one-half of the loads on the interior nodes since only one element contributes to the load.

In Figure 6.6, there are nine interior nodes and two corner nodes at each end of the panel. The load on each of the interior nodes is equal to one unit. The loads on the corner nodes are equal to one-half of a unit.

STEP 6—SOLVING FOR NODAL DISPLACEMENTS

The nodal displacements for the unrestrained nodes are found by solving Equation 6.18. That is to say, the displacements are found by inverting the restrained global stiffness matrix to give the following:

$$\{d_1\} = [K_{11}]^{-1} \{F_1\} \tag{6.20}$$

where $\{d\} = \begin{Bmatrix} u \\ v \end{Bmatrix}$

u, the displacements in the x direction

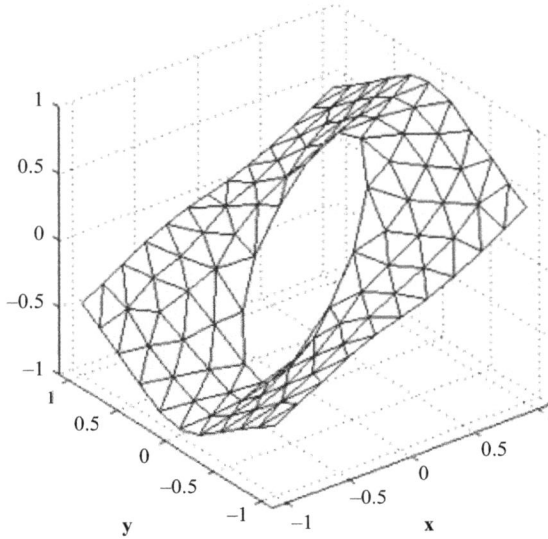

Figure 6.7. Displacements due to the end loads.

As noted earlier in Equation 6.12, the nodal displacements in the x direction, the u's, are listed first in the displacement vector and the displacements in the y direction, the v's, are listed second. The displacements in the x direction for the problem shown in Figure 6.6 are presented in Figure 6.7.

As can be seen in Figure 6.7, the u displacements at the two ends of the panel behave as would be expected. The magnitudes of the displacements on the top and bottom edges are smaller than the displacements that are closer to the hole. This occurs because the panel is *softer* in these regions because of the presence of the cutout.

STEP 7—EXTRACTION OF ELEMENTAL STRAINS AND STRESSES

The elemental strains are extracted with equations that are used to develop the element stiffness matrix. The elemental strains are given in terms of the strain gradient quantities in Equation 6.6, which is reproduced here as follows:

$$\{\varepsilon\} = [\mathrm{T}]\{\varepsilon,\} \tag{6.21}$$

where

$$\{\varepsilon\}^T = [\varepsilon_x \ \varepsilon_y \ \gamma_{xy}]$$

$$\{\varepsilon,\}^T = [(u_{rb})_0 \ (v_{rb})_0 \ (r_{rb})_0 \ (\varepsilon_x)_0 \ (\varepsilon_y)_0 \ (\gamma_{xy})_0]$$

$$[T] = [[T_0] \ [T_\varepsilon]]$$

Since the finite element model is solved for nodal displacements and the strains in Equation 6.21 are expressed in terms of strain gradient quantities, we must use the transformation from the strain gradient to nodal displacements in order to find the elemental strains. This is accomplished by applying the transformation given by Equation 6.13 that is reproduced here for convenience:

$$\{\varepsilon,\} = [\Phi]^{-1}\{d\} \tag{6.22}$$

When Equations 6.21 and 6.22 are combined, we have the following expression for computing the elemental strains:

$$\{\varepsilon\} = [T][\Phi]^{-1}\{d\} \tag{6.23}$$

With the availability of the elemental strains, the elemental stresses are found by applying the constitutive relation that was defined in Equation 6.7 during the element formulation process. The stress–strain relationship is the following:

$$\begin{Bmatrix} \sigma_x \\ \sigma_y \\ \tau_{xy} \end{Bmatrix} = [E] \begin{Bmatrix} \varepsilon_x \\ \varepsilon_y \\ \gamma_{xy} \end{Bmatrix} \tag{6.24}$$

We will now present samples of the stresses and strains produced for one mesh. In the next section, the stresses and strain produced by the finite element models formed with the refined meshes will be presented and compared in order to identify some characteristics of finite element solutions.

6.6 PRESENTATION OF THE POINTWISE AND ELEMENTAL STRESSES AND STRAINS

The stress and strain distributions for planar problems can be seen most clearly in three dimensions when they are plotted as pointwise quantities.

If the distributions are plotted with *elemental quantities*, the plots are difficult to interpret because of the discontinuities that exist as the interelement jumps in the stresses and strains. In other words, the use of pointwise quantities smooth the plots.

In this presentation, the pointwise stresses and strains are computed as an average of the elemental quantities that intersect at a node. As an example, let us consider the nodal strain at node 21, which is in the lower left-hand corner in Figure 6.4a. The following six elements in Figure 6.4b can be seen to intersect at node 21: 2, 8, 61, 59, 33, 70, and 69. The nodal strains at node 21 are taken as the average of these quantities for the individual elements.

These averaged nodal quantities will be referred to as *smoothed quantities or smoothed results* in the text that follows. When the pointwise stresses are computed in this way, they can differ widely from the elemental results. Later, when we compute error estimates, the differences between the smoothed quantities and the elemental quantities will be put to good use in estimating the errors in the finite element models.

The nodal values for the strain component ε_x are presented in Figure 6.8a. As can be seen, the strains at two locations on the interior circle are significantly higher than the strains elsewhere on the panel. The strain distributions in the neighborhood of these critical regions are increasing rapidly. As we will see, it is the existence of large gradients in the strains that produces significant errors in a finite element result. This is the case because the limited strain modeling capability of the individual finite elements cannot capture the actual strain distribution.

The nodal stresses for the stress component σ_x are shown in Figure 6.8b. The critical points contained in Figure 6.8a are seen as stress concentrations in Figure 6.8b. The maximum critical stress value for the nodal values is found to be 12.95 units/units of area.

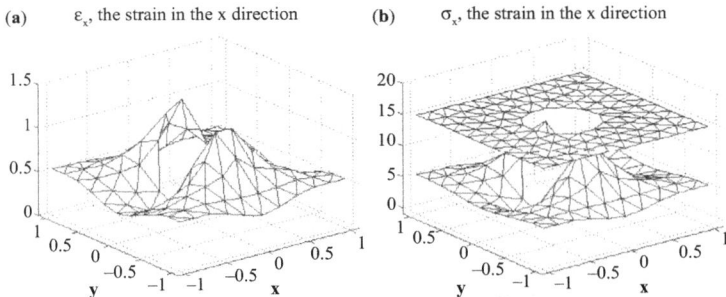

(a) ε_x, the strain in the x direction   (b) σ_x, the strain in the x direction

Figure 6.8. Nodal strains and stresses for normal values in the x direction, $h0 = 0.2$: (a) nodal strain values for ε_x and (b) nodal stress values for σ_x.

The plane shown in Figure 6.8b with a stress value of 15 units/unit area represents the magnitude of the stress concentration for a panel of infinite length and width (Budynas 1999). This plane contains an internal circular hole with a uniformly distributed load of 5 units/unit of area on the boundary at x = ±infinity. The actual stress concentration for the finite panel pictured here will be significantly higher than that for the infinite panel with an identical loading condition. This is the case because a smaller area is carrying the load at the critical points.

The preceding comparison with the maximum stress for the infinite panel is included to show that the nodal results for the stress concentrations in this rather coarse mesh are not very accurate. The maximum stress value for the finite panel should be higher than the stress concentration for the infinite panel. As can be seen, the finite element model produced maximum stresses that are less than those for the infinite panel. As we will see, the mesh must be refined in order to get closer to the exact value for this finite problem. For the problem pictured in Figure 6.8, the nominal size of the mesh is defined by the variable h0 in the mesh generation program. In this case, h0 = 0.2 units.

As mentioned earlier, a plot of the elemental stresses and strains over the total domain of a problem is difficult to visualize because of the discontinuities in the stresses. However, when the elemental stresses are plotted on the boundary of the internal hole and on the top boundary of the panel, salient characteristics of a finite element result are clearly seen. The three stress components on the two boundaries for the coarse mesh shown in Figures 6.3 to 6.8 will be presented in turn.

For this case, the elements on the boundary of the interior hole are identified in Figure 6.9. The locations of these elements in the overall mesh can be seen by comparing Figure 6.9 with Figure 6.4b.

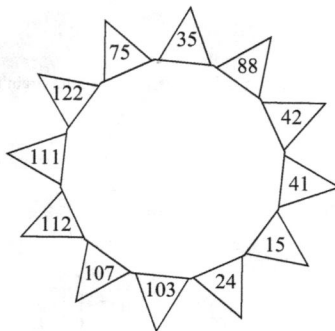

Figure 6.9. Elements on the hole boundary.

In order to get the three normal and tangential values for a given element, we must transform the stresses in the x–y coordinate system to a coordinate system aligned with the edge of the individual elements. This requires a rotation of the coordinates through an angle θ, the angle from the positive x axis to the center of the element. The transformed strains are computed with the following stress transformation:

$$\left\{\begin{matrix} \sigma_{x'} \\ \sigma_{y'} \\ \tau_{x'y'} \end{matrix}\right\} = \begin{bmatrix} \cos^2\theta & \sin^2\theta & +2\cos\theta\sin\theta \\ \sin^2\theta & \cos^2\theta & -2\cos\theta\sin\theta \\ -\cos\theta\sin\theta & \cos\theta\sin\theta & \cos^2\theta-\sin^2\theta \end{bmatrix} \left\{\begin{matrix} \sigma_x \\ \sigma_y \\ \tau_{xy} \end{matrix}\right\} \quad (6.25)$$

The three components of the stress on the boundary of the internal circle are shown in Figure 6.10. The critical tangential normal stress is shown in Figure 6.10a. As can be seen, two stress concentrations are present. They correspond to locations at the top and bottom of the hole shown in Figures 6.4 and 6.9. The maximum elemental stress value on the boundary is approximately 17.55 unit/unit of area. The normal stress components

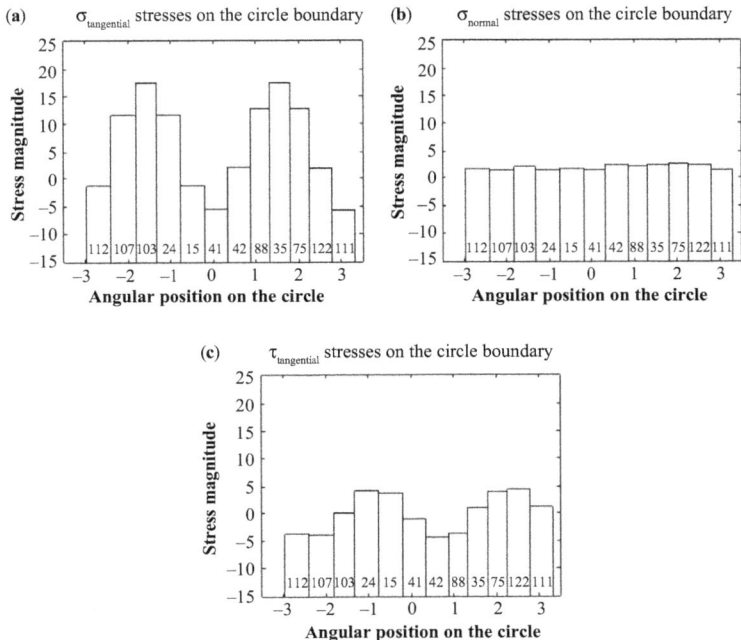

Figure 6.10. Stress components on the surface of the circular hole: (a) tangential normal stress, (b) normal stress normal to hole, and (c) tangential shear stress.

that are perpendicular to the surface of the individual elements on the circle are shown in Figure 6.10b. The shear stress components on the faces of the individual elements that define the circle are shown in Figure 6.10c.

The basic nature of a typical finite element result can be seen in the three stress components of Figure 6.10. Interelement jumps in the stresses exist between the elements. Later, we will see that the magnitudes of these jumps are indicative of the level of the modeling error *at this location* in the finite element approximation. In other words, the interelement jumps provide the basis for identifying the locations and magnitudes of the errors in a finite element representation. The model can then be refined in regions of excessive error in order to improve the accuracy of the results.

The interelement jumps exist when an element cannot represent the higher-order strain gradient conditions that are present in the exact solution on the domain of the element. These errors are called *discretization errors* because they are caused when a discrete representation cannot capture the continuous solution. As we will see in a later chapter, the refinement guides are based on the fact that the strain gradient notation is expressed in terms of the physical quantities that produce the displacements in the continuum, namely, rigid body motions and strain quantities. As a result, this notation can be used to quantify the inability of an individual finite element to capture the exact solution.

Another characteristic of a finite element solution is seen in Figures 6.10b and 6.10c. In this model, these two components of stress should be equal to zero because the interior of the circle is a free boundary. In other words, the internal boundary is not loaded in this problem, so the normal and shear stresses on this boundary should be equal to zero.

The existence of errors in the boundary stresses is a characteristic of an inaccurate finite element solution. This is the case because the finite element method is a variational approach. Variational formulations are often referred to as *weak formulations*. This designation is used because only the displacement boundary conditions must be specified in variational formulations of plane stress. The stress boundary conditions are not specified in a variational formulation so the finite element representations only approach the actual result as the model is improved.

In contrast, *finite difference models* are referred to as *strong formulations*. This is the case because both displacement and boundary stresses conditions must be specified in the model. In other words, the finite element and the finite difference approximations are different types of solution techniques. We will use this distinction in the development of error measures in the next chapter.

Finally, when the smoothed stresses, which are formed from the average of the nodal stresses, are superimposed on the stresses in the individual elements on the boundary of the cutout, the results are shown in Figure 6.11. As can be seen in Figure 6.11a, the two stress representations for the critical stress component vary widely.

When the maximum value of σ_x for the smoothed solution is compared to the maximum value for the elemental value for σ_x, there is a significant difference between the two values. When the two values are compared in an error calculation, $((17.38 - 12.95)/17.38) \times 100$, we get a difference of 25.50 percent for the problem shown in Figure 6.4. In the next chapter, we will extend the idea of comparing the two types of representations to form a pointwise error estimator.

In the case of the two stress components on the boundary of the cutout, they do not vary widely but they also do not satisfy the boundary condition on the interior of the circle. The normal stress on the boundary

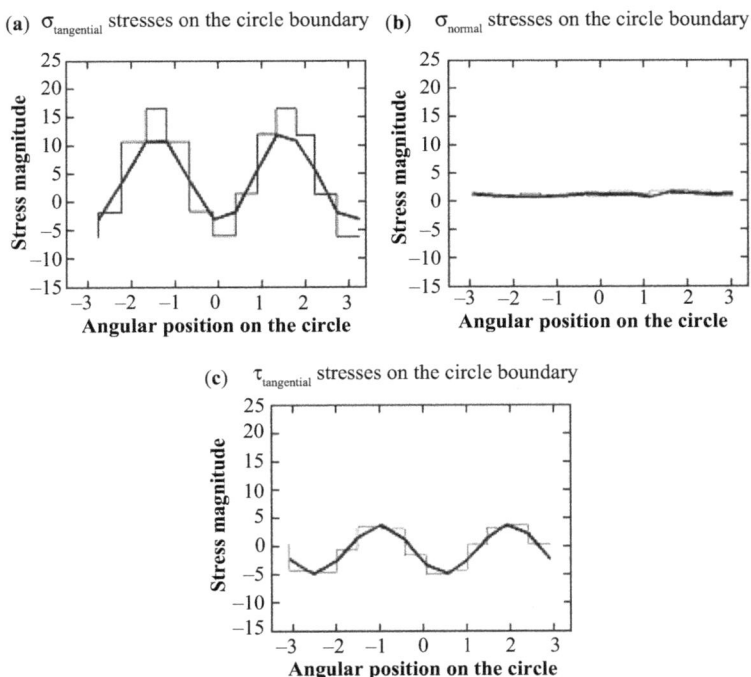

(a) $\sigma_{tangential}$ stresses on the circle boundary (b) σ_{normal} stresses on the circle boundary

(c) $\tau_{tangential}$ stresses on the circle boundary

Figure 6.11. Comparison of stress components on the surface of the circular hole: (a) tangential normal stresses, (b) perpendicular normal stresses, and (c) tangential shear stresses.

shown in Figure 6.11b, which should be zero, is nearly constant at approximately 2 units. In contrast, the shear stress on the boundary of the cutout
varies sinusoidally between plus +5 units and −5 units. It, too, should be
equal to zero.

In contrast to the *inaccurate* representation of the exact solution by
the finite element model on the interior of the cutout for this model, the
finite element representation on the top boundary is *nearly perfect*. In this
case, the smoothed and elemental stresses for the three stress components
essentially coincide with each other. Furthermore, the interelement jumps
are practically nonexistent. The smoothed stresses and the associated
nodal points are presented in Figure 6.12.

The normal stress in the x direction presented in Figure 6.12a is nearly
constant at 5 units per square unit. This is equal to the stress that is applied
to the panel at the two ends. In other words, the top edge is far enough
from the discontinuity in the panel that it is exhibiting the load that would
exist if the circular cutout was not present. Similarly, the normal and shear
boundary stresses shown in Figures 6.12a and 6.12b are nearly equal to

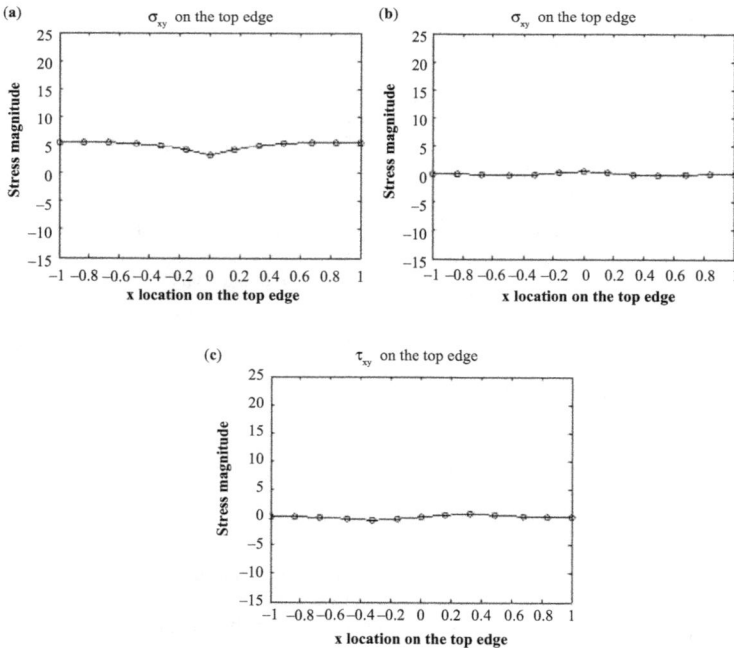

Figure 6.12. Stress components on the top boundary of the panel: (a) normal
stress in the x direction, (b) normal stress in the y direction, and (c) shear stress.

zero along the full length of the panel. This is close to the result that would exist in the exact solution.

The accuracy of the stresses on the top boundary for this coarse mesh contrasts to the inaccurate results for the stresses on the internal hole. The difference is easy to explain. The stresses on the top edge are relatively constant. This means that the rate of change of the stresses on the top edge of the panel is close to zero. The finite element model is formed with three-node elements that are only capable of representing constant strains. Since these elements are representing strain states that they are capable of representing, they represent them well.

In contrast, the strain distributions on the boundary of the cutout are rapidly changing. This occurs because of the very nature of a stress concentration. The stress concentration is a high point that is surrounded by rapidly changing stresses. The stresses oscillate from a relatively low value to a maximum value and go down again.

Since a single three-node element can only represent constant values, the finite element model must present these changes with stepwise discontinuities. As a result, this attempt to model a rapidly changing continuous result with a finite number of low-order elements produces discontinuities. Consequently, the only way to improve the results with a model formed with three-node elements is to increase the number of elements in regions of rapidly changing strains.

As we will see in the next section, the errors in the finite element model in regions of rapidly changing strains are reduced as the mesh is refined. The interelement jumps are smaller, the boundary stresses get closer to zero, and the differences between the smoothed and the elemental stresses decrease. These improvements presage the fact that these three characteristics, in conjunction with the *finite* strain modeling characteristics of individual elements, will be exploited to create error estimators and refinement guides in later chapters.

6.7   THE EFFECTS OF MESH REFINEMENT

As just noted, the maximum stress value for the finite element model shown in Figure 6.4 is significantly different than the expected value for the less severe stress concentrations in an infinite panel. When the mesh size is halved by setting to h0 = 0.1 in the mesh generator, the resulting mesh is shown in Figure 6.13. Note that the total load for this case is the same as that for the coarser mesh shown in Figure 6.6.

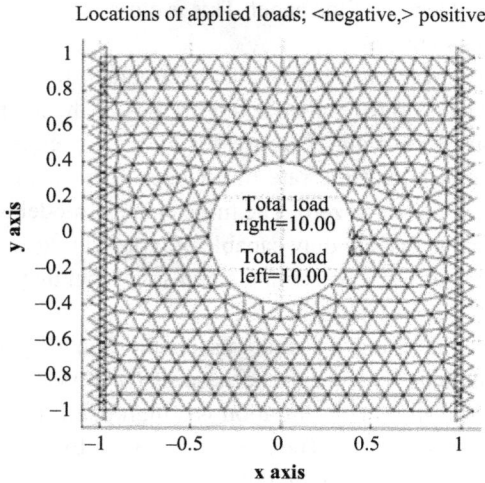

Figure 6.13. Mesh with applied loads identified.

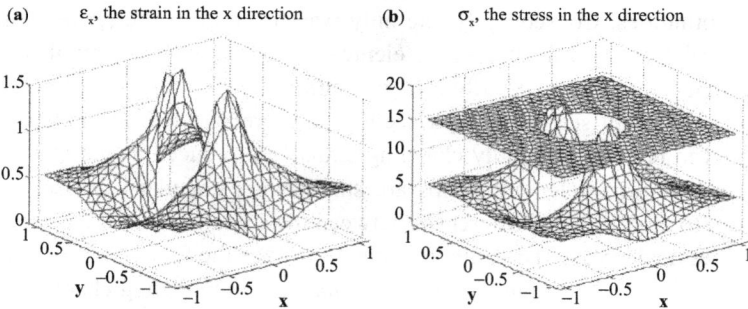

Figure 6.14. Nodal strains and stresses for normal values in the x direction, $h0 = 0.1$: (a) nodal strain values for ε_x and (b) nodal stress values for σ_x.

The pointwise strains and the stresses on the domain of the problem produced by this refined mesh are presented in Figure 6.14. As can be seen, the model has been improved. In fact, the maximum pointwise stress of 17.21 units/unit area exceeds the 15 units/unit area that exists in an infinite panel.

The elemental stresses on the boundary of the cutout are shown in Figure 6.15. As can be seen, the maximum critical stress has increased to 20.93 from 17.38 in the previous case, and the boundary stresses are closer to zero. Note that the improvements in the three elemental stress results are accompanied by a reduction in the size of the interelement jumps.

(a) $\sigma_{\text{tangential}}$ stresses on the circle boundary (b) σ_{normal} stresses on the circle boundary

(c) $\tau_{\text{tangential}}$ stresses on the circle boundary

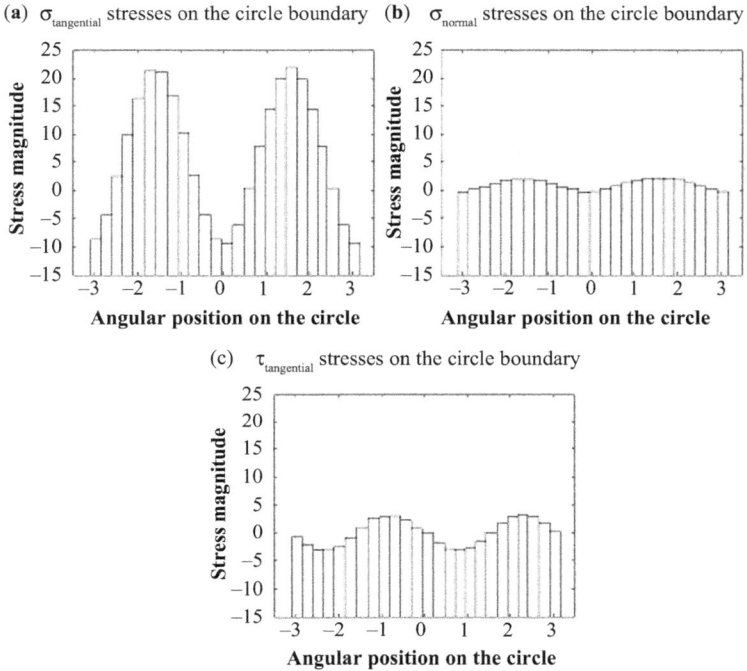

Figure 6.15. Stress components on the boundary of the circular hole: (a) tangential normal stress, (b) perpendicular normal stress, and (c) tangential shear stress.

The smoothed stresses on the cutout are shown superimposed on the elemental stresses in Figure 6.16. As can be seen when Figure 6.16 is compared to Figure 6.11, the smoothed stresses are closer to the elemental stresses than they were in the previous model.

In the previous mesh, we saw that the elemental representation of the critical stress was significantly different than the smoothed representation. The difference between the maximum stress values was 25.50 percent. In this case, a significant difference still exists between the two maximum stresses, but it is smaller than in the previous case. The maximum elemental stress for the critical point shown in Figure 6.15a is equal to 20.57 units/unit area. When the smoothed and elemental values are compared in an error computation, $((20.93 - 17.21)/20.93)*100$, we get a difference of 17.80 percent. Again, note that the smaller differences between the smoothed and elemental stresses are accompanied by smaller inter-elements jumps in the stresses.

When we compare the three components of the boundary stress shown in Figure 6.15 with those contained in Figure 6.10, we have direct

(a) $\sigma_{tangential}$ stresses on the circle boundary (b) $\sigma_{tangential}$ stresses on the circle boundary

(c) $\sigma_{tangential}$ stresses on the circle boundary

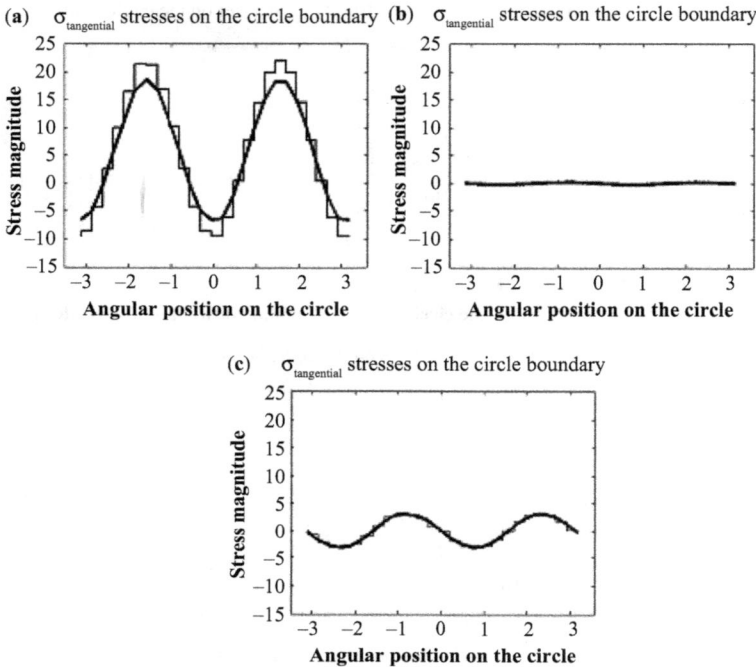

Figure 6.16. Comparison of stress components on the surface of the circular hole: (a) tangential normal stresses, (b) perpendicular normal stresses, and (c) tangential shear stresses.

visual evidence that the mesh refinement has improved the finite element representation. First, the magnitudes of the interelement jumps in all three stress components have been reduced. Second, the boundary stresses shown in Figures 6.13b and 6.13c are closer to zero than those shown in Figures 6.15b and 6.15c. That is to say, the boundary stresses are better represented in the refined model.

We could continue to *uniformly refine* the mesh until the stress and strain results converge indicating that we have arrived at an acceptable approximation of the actual solution. However, this strategy is both an impractical and unnecessary way to reach an acceptable solution. It is impractical because of the large number of elements that would exist in the model. It is unnecessary because the majority of the new elements would be located in regions of low error.

The problem with uniform refinement is demonstrated by Figure 6.12. The results for the stresses on the top boundary produced by this coarse mesh are almost perfect. When this course mesh was uniformly refined, the stress results on the top boundary were, at most, imperceptibly improved. That is to say, the addition of new elements on this boundary did not improve the result. They were unnecessary.

The majority of the elements added in a uniform refinement are not needed because the strains that occur away from the stress concentration do not change rapidly. As a result, these regions are accurately represented with fewer elements. *This fact is the driving force and motivation for developing error estimators and mesh refinement guides.* With these two capabilities, the elements that are added to the model are placed only where they are needed in order to improve the model.

The impracticality of uniform refinement exhibits itself in the next example. In order to represent the regions with critical points with very small elements, the *nonuniform mesh* shown in Figure 6.17a had to be used. The model with a uniform mesh exceeded the memory of the computer used to develop these examples.

In this example, the nominal size of the elements on the boundary of the cutout is specified as h0 = 0.005 units in the mesh generator. This

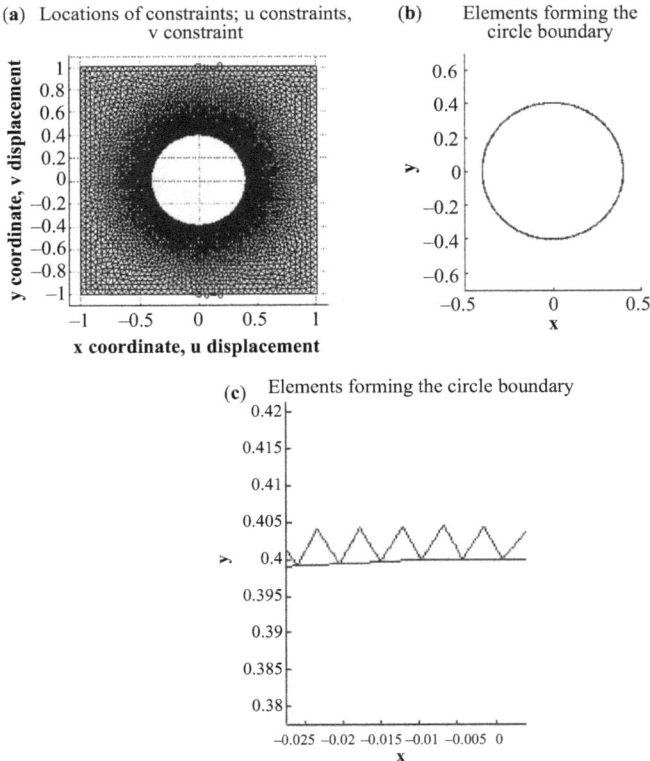

Figure 6.17. A highly refined nonuniform finite element mesh: (a) fine, graduated mesh; (b) individual boundary elements; and (c) close-up of boundary elements.

is 20 times smaller than the nominal element size of h0 = 0.1 used in the previous example. The nonuniform mesh for this problem is shown in Figure 6.17a. It contains 17,785 elements and has 9,208 nodes.

The elements that form the boundary of the internal circle are shown in Figure 6.17b. There are 486 elements on this boundary. As can be seen in both Figures 6.17a and 6.17b, the elements on this boundary are too small to be distinguished separately. In Figure 6.17c, a portion of the boundary is expanded so the individual elements can be seen.

The critical stress is presented in Figure 6.18. As can be seen in Figure 6.18a, the elements are so small that it is impossible to distinguish

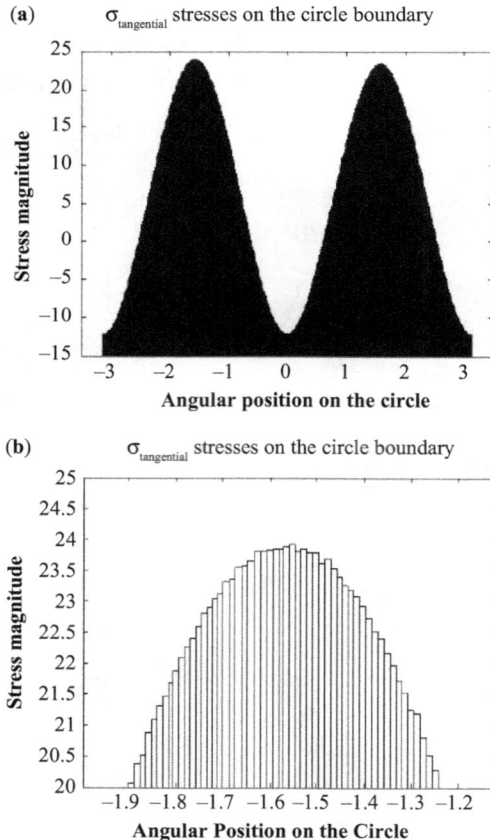

Figure 6.18. Critical stress component on the surface of the circular hole: (a) tangential normal stress and (b) an expanded portion Figure 6.18a.

individual elements. In Figure 6.18b, an expanded view of the stress concentrations at the top of the circle is shown. The maximum value is 23.92. This is significantly higher than the critical value for σ_x of 20.93 found in the previous model. In addition, the interelement jumps in this figure are significantly smaller than in the previous models.

The smoothed stresses on the cutout are shown superimposed on the elemental stresses in Figure 6.19. As can be seen, the smoothed stresses and the elemental stresses are nearly identical for all three stress components. As we will see later, this indicates that the finite element model is a good representation of the continuous problem.

Finally, when the maximums of the smoothed and elemental values of σ_x are compared with an error computation, we get the following: $((23.98 - 23.68)/23.98) \times 100 = 1.23$ percent. Since the two previous models produced error computations of 23.49 percent and 17.80 percent, respectively, this heavily refined model produced a significantly smaller difference.

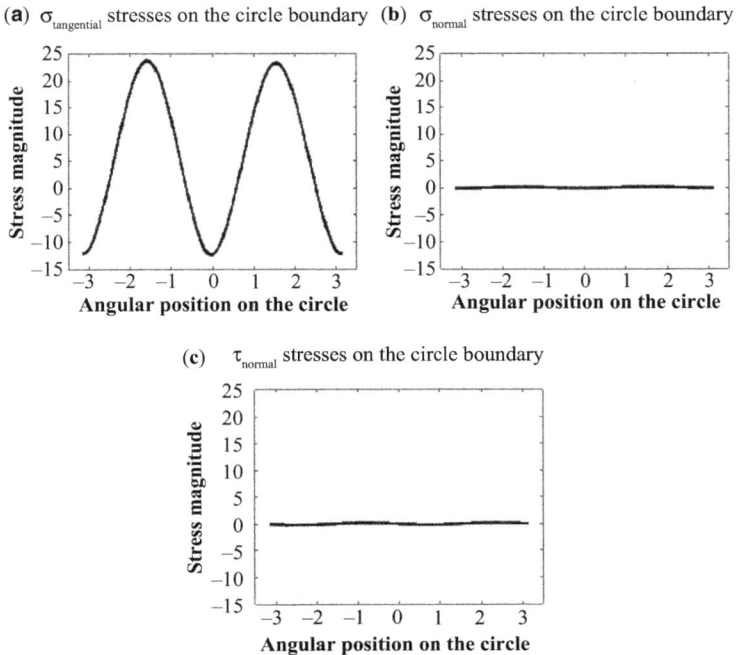

(a) $\sigma_{\text{tangential}}$ stresses on the circle boundary   **(b)** σ_{normal} stresses on the circle boundary

(c) τ_{normal} stresses on the circle boundary

Figure 6.19. Comparison of stress components on the surface of the circular hole: (a) tangential normal stresses, (b) perpendicular normal stresses, and (c) tangential shear stresses.

The boundary stresses that should be equal to zero are shown in Figure 6.19b and 6.19c. When these plots are compared to the equivalent stress components in Figures 6.11 and 6.15, we see that they are significantly improved.

As noted earlier, this problem, with a nominal element size on the interior boundary of h0 = 0.005, is the largest problem that could be solved with the computer being used. When the problem was solved with h0 = 0.0055, the difference between the maximum elemental stresses for the two problems was 0.25 percent. As a result of this relative convergence, we can conclude that the error computation that compares the elemental stresses to the pointwise stresses is indicative of the accuracy of the solution. In other words, the estimate of 1.23 percent was a reasonable estimate of the accuracy of the representation of the critical stresses.

6.8 SUMMARY AND CONCLUSIONS

In addition to presenting a compact tutorial for forming and solving finite element models, this chapter provided the intuitive basis for creating the error estimators and the refinement guides that will be presented in the next two chapters. The causes of the errors in finite element solutions and their effects on the solutions are identified and demonstrated.

It was seen that discontinuities are produced in the stress and strain representations when an individual finite element cannot represent the complexity of the strain distribution that exists on the portion of the domain it is representing. Another form of discontinuity was seen on the loaded boundaries of a finite element model. In the examples presented, we saw that the stresses on the free boundaries were not equal to zero as should be the case. However, they approached zero as the model was improved.

When a mesh is refined, each of the elements, by definition, represents a smaller portion of the overall problem. Consequently, the complexity of the strain distribution over the domain of an individual element is reduced in critical regions where there are high strain gradients. As a result, the smaller element is better able to represent the strain distribution on its domain with its limited modeling capability. Since the individual elements are better able to represent the actual strain distribution, the interelement jumps between the elements are reduced. This, in turn, causes the smoothed stresses to approach the elemental stress results as the mesh is refined because of the smaller differences between the adjacent elements.

In the next chapter, a solid theoretical basis for the convergence of the smoothed and discontinuous solutions is presented. Then, pointwise error estimators are developed in terms of the differences between the two types of solutions.

6.9 EXERCISES

1. Use the contents of Appendix 6A to form the stiffness matrix for two elements of the same shape with different sizes. Note that the stiffness matrices are identical.
2. Compute the eigenvalues for a series of elements with different shapes and compare their condition numbers to an equilateral triangle. The eigenvalue function can be found in MATLAB through the use of "help eig."
3. Create a program using the outline presented in this chapter and the mesh generator presented in Chapter 5 and the three-node element presented in Appendix 6A to form a finite element model for the Kirsch problem.

APPENDIX 6A SYMBOLIC MATLAB FORMULATION OF A THREE-NODE ELEMENT STIFFNESS MATRIX

6A.1 INTRODUCTION

This appendix uses symbol manipulation to form the stiffness matrix for a three-node finite element representing plane stress. This presentation is made for three reasons:

1. To eliminate any ambiguity about the formulation of a stiffness matrix with strain gradient notation by relating the equations in the main text to MATLAB code and the results produced by these MATLAB operations.
2. To demonstrate the reason and to reinforce the fact that the strain gradient approach requires only a small number of integrals to be evaluated.
3. To demonstrate how operations can be *precomputed* symbolically so that the same calculation does not have to be repeated for every stiffness matrix that is formed.

6A.2 MATLAB DRIVER FOR THREE-NODE STIFFNESS MATRIX FUNCTION

1. clc, format compact, clear all, close all % These operations guarantee a clear machine.
2. display('dbtype E:\AAChapt6\ThreeNodeDriver') % If executed, the code is listed with line numbers.
3. %
4. % This program drives the function that forms the 3-node stiffness matrix that
5. % uses the output of a symbol manipulation program. This means that the computations
6. % are reduced because of observations made during the element formulation.
7. %
8. % Input Data: The physical parameters and the geometry for several elements.
9. %
10. % Enter physical quantities:
11. E=10.0 % Young's modulus.
12. nu=0.3 % Poisson's ratio.
13. t=1 % Thickness.
14. %
15. % Geometry for Triangle 1
16. % x=[0.0 2.0, 1.0], y=[0.0, 0.0, 1.0]
17. %
18. % Geometry for Triangle 2
19. % x(1)=0.0,x(2)=4.0,x(3)=2.0, y(1)=0.0,y(2)=0.0,y(3)=2.0
20. %
21. % Geometry for Triangle 3
22. % x(1)=-1.0,x(2)=1.0,x(3)=0.0, y(1)=0.0,y(2)=0.0,y(3)=1.0
23. %
24. % Geometry for Triangle 4, one-half triangle 3. Get same K.
25. %x=[-0.5 0.5 0.0], y=[0.0 0.0 0.5]
26. %
27. % Geometry for Triangle 5, Equilateral
28. x(1)=-1.0; x(2)=1.0; x(3)=0.0, y(1)=0.0;,y(2)=0.0;,y(3)=0.866*2
29. %
30. figure(1)

31. plot([x(1,:),x(1,1)],[y(1,:),y(1,1)])
32. %
33. [K]=SymStiff3(x,y,E,nu,t)

6A.3 MATLAB FUNCTION FOR FORMING THREE-NODE STIFFNESS MATRIX

1. function [K]=SymStiff3(x,y,E,nu,t)
2. % Form stiffness matrix for 3-node element using output of symbol manipulation program.
3. display('dbtype E:\AAChapt6\SymStiff3')
4. %
5. % Input data
6. % x's = x locations of element nodes.
7. % y's = y locations of element nodes.
8. % E = Young's modulus.
9. % nu = Poisson's ratio.
10. % t = thickness of element.
11. %
12. % Internally generated parameters.
13. % a = Coefficient in the plane stress constitutive relation, a = (1-n)/2.
14. % m = multiplier in the plane stress constitutive relation, m = E/(1-n^2).
15. % I1 = The single integral in the 3 node stiffness matrix, the area.
16. %
17. x1=x(1);x2=x(2);x3=x(3); y1=y(1);y2=y(2);y3=y(3); % Nodal coordinates.
18. %
19. n=nu, a=(1-n)/2, m=E/(1-n^2); % Physical quantities.
20. I1=(x1*y2-x2*y1-x1*y3+x3*y1+x2*y3-x3*y2)/2; % Area of triangle.
21. %
22. U22 =(I1*m*t)*[1, n, 0; n, 1, 0; 0, 0, a]; % Constitutive Relation.
23. %

24. PhiInv21 =[y2-y3,y3-y1,y1-y2;0,0,0;x3-x2,x1-x3,x2-x1]/(2*I1);
    % Eq. 6.13
25. PhiInv22 =[0,0,0;x3-x2,x1-x3,x2-x1;y2-y3,y3-y1,y1-y2]/(2*I1);
    % Eq. 6.13
26. PhiInvTran12 =[y2-y3,0,x3-x2;y3-y1,0,x1-x3;y1-y2,0,x2-x1]/
    (2*I1); % Eq. 6.16
27. PhiInvTran22 =[0,x3-x2,y2-y3;0,x1-x3,y3-y1;0,x2-x1,y1-y2]/
    (2*I1); % Eq. 6.16
28. %
29. % Now form the partitions that make up Eq. 6.16.
30. KPart11=PhiInvTran12*U22*PhiInv21; % Form the partition
    K11 for Eq. 6.16.
31. %
32. KPart12=PhiInvTran12*U22*PhiInv22; % Form the partition
    K12 for Eq. 6.16.
33. %
34. KPart22=PhiInvTran22*U22*PhiInv22; % Form the partition
    K22 for Eq. 6.16.
35. %
36. % Now assemble the full matrix, Eq. 6.16
37. K=[KPart11 KPart12; KPart12' KPart22]

CHAPTER 7

POINTWISE ERROR ESTIMATORS

7.1 INTRODUCTION

The *motivation, insight,* and *approach* for creating elemental error estimators were provided by the results presented in the previous chapter. *The motivation* for developing the ability to estimate errors in individual elements was demonstrated when uniform refinement was applied to the Kirsch problem shown in Figure 7.1.

It was shown that the strategy for improving a model by simply reducing the size of every element is both unnecessary and impractical. Uniform refinement is *unnecessary* because a majority of the elements are added in regions that already have acceptable accuracy. Repeated application of the process is *impractical* because adding unneeded elements produces overly large and, hence, inefficient models.

A comparison of Figures 6.10a, 6.15a, and 6.18b reproduced here as Figure 7.2 provides *the insight* that allows the *source of the errors* that exist in finite element results to be identified. As can be seen, the magnitudes and the accuracy of the critical stresses in the representations are increasing as the size of the elements is reduced. Correlated with the improvement in the solution is a reduction in the magnitudes of the interelement jumps in the stress. Thus, we can conclude that the jumps quantify the inability of the individual elements to represent the exact solution.

The relationship between the errors and the jumps can be explained as follows. As can be seen in Figure 7.2, two things occur when the size of an element is reduced. On the one hand, an individual element covers a smaller portion of the domain of the problem. On the other hand, the sizes of the interelement jumps decrease as the element size decreases.

In other words, the portion of the solution that a smaller element must represent in a region of high error is simpler than that for a larger element. In terms of a Taylor series representation, the lower order coefficients in this smaller portion of the exact solution contribute more to the representation than they did in the larger segment of the exact solution. As a result,

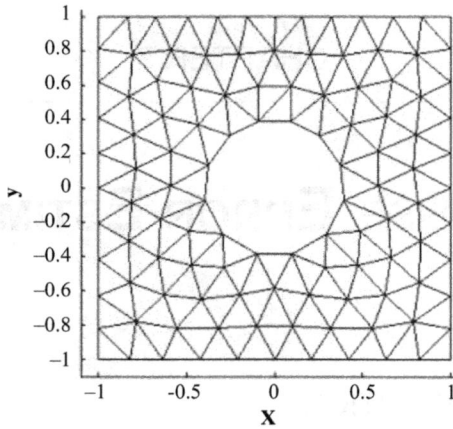

Figure 7.1. Finite element mesh.

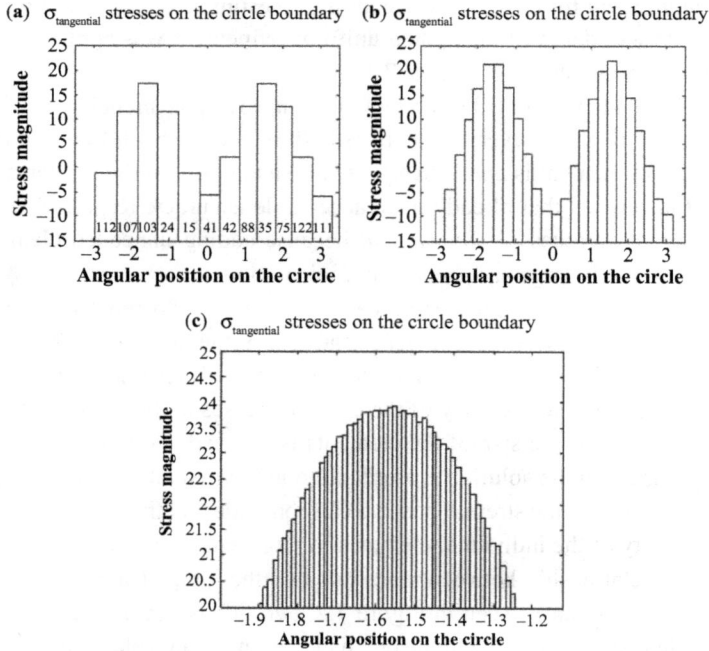

(a) $\sigma_{tangential}$ stresses on the circle boundary (b) $\sigma_{tangential}$ stresses on the circle boundary

(c) $\sigma_{tangential}$ stresses on the circle boundary

Figure 7.2. Tangential normal stress distributions: (a) nominal mesh size, h0 = 0.2; (b) nominal mesh size, h0 = 0.1; and (c) nominal mesh size, h0 = 0.005.

the low-order modeling capabilities of an individual element are better able to represent this smaller portion of the exact solution. Hence, the errors and the interelement jumps are smaller.

In addition to providing the insight for identifying the source of the errors in finite element models, *an approach* for estimating the magnitude of the errors in individual elements was demonstrated in Chapter 6. In that demonstration, a smoothed solution was formed by averaging the elemental stresses at the nodes. The errors were estimated in terms of the difference between the discontinuous finite element stresses and the *smoothed stress representations* at the nodes. The use of this difference in stresses as an error measure is useful because it allows the errors to be expressed in terms of quantities that are important in the design process, namely, stresses or strains. However, the use of this difference between solutions as an error estimator was not put on a solid theoretical foundation.

One might argue that the smoothed solution is closer to the exact solution of the problem being solved than the finite element result because it does not contain discontinuities. This is an appealing argument because the size of interelement jumps is an indicator of the magnitude of the errors. However, this conjecture is difficult to support.

One reason for this difficulty is that the smoothed stresses in intermediate refinements are often less accurate than the discontinuous finite element result. This is illustrated by comparing the initial results shown in Figure 6.11a to the nearly converged results shown in Figure 6.18a, which are presented here as Figure 7.3a and 7.3b, respectively. When these results are compared, we see that the smoothed result at the point of maximum stress in the initial model is further from the nearly converged result than is the discontinuous finite element result.

Since we cannot consider the smoothed solution as a better solution than the finite element result, the following question arises: *How are we going to guarantee that the difference between the two types of solution can serve as a reliable error estimator?* We will provide this guarantee

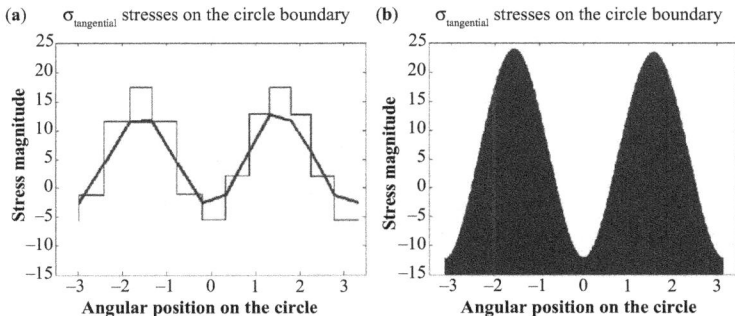

Figure 7.3. Comparison of approximate solutions: (a) finite element and smoothed approximations and (b) nearly converged finite element result.

by showing that the smoothed solution must also converge to the exact solution. Thus, the differences between the two distributions can be interpreted as deficiencies in the model. These differences will be quantified to give the error estimates.

7.2 OBJECTIVES

The primary objective of this chapter is to provide a solid theoretical basis for using the difference between the smoothed solution and the discontinuous finite element result as the basis for an error estimator. The secondary objective is to demonstrate the efficacy of using the smoothed solution in an error estimator.

The argument for using the difference between these two significantly different types of solution as the basis for an error estimator is expressed succinctly by Sherlock Holmes. When speaking to his associate Dr. John Watson, he says,

> When you follow two separate chains of thought, Watson, you will find some point of intersection that should approximate the truth. (Doyle 1917)

By definition, the discontinuous finite element solution is a *product* of the finite element method. In contrast, the smoothed solution is an *approximation* of a finite difference solution. As we shall see, both solution techniques produce results that will converge to exact solutions by minimizing the potential energy in the problem. However, the two methods use significantly different approaches to minimize the potential energy and produce their solutions. These differences are described in detail in the next section.

In other words, both methods pursue the same goal, but with *different chains of thought*. Thus, we shall conclude that the differences between the two approximate solutions are due to errors in the model that approximates the problem since they are derived from the same mesh. The effectiveness of using the differences between the two solutions to identify error in finite element models is demonstrated in a later section.

In brief, the difference between the two approximate solution techniques for minimizing the potential energy in the problem can be described as follows. The finite element method *minimizes the strain energy in a model of the physical system*. In contrast, the finite difference method models *the governing differential equations* and *the boundary conditions* at the nodal points. These equations supply the necessary and sufficient conditions for minimizing the potential energy.

As a final note, it should be emphasized that the approximate finite difference solution used in the error analysis is derived from the finite element result. *A finite difference model does not have to be formed and solved.* This note is added because of a misapprehension that has occurred in several seminars and paper presentations made by the author.

7.3 A COMPARISON OF THE FINITE ELEMENT AND FINITE DIFFERENCE METHODS

In this section, the similarities and the differences in finite difference and finite element methods are presented in detail. The two methods are shown schematically in Figure 7.4. As can be seen, the starting point for the two approximate solution techniques is identical. The objective of both methods is to find the displacements that minimize the strain energy in their respective models.

Furthermore, if the models have the same nodal locations, they will converge to the same result as the model is refined. However, the fact that they find these displacements in different ways is implied by the different paths the two methods take to reach their results.

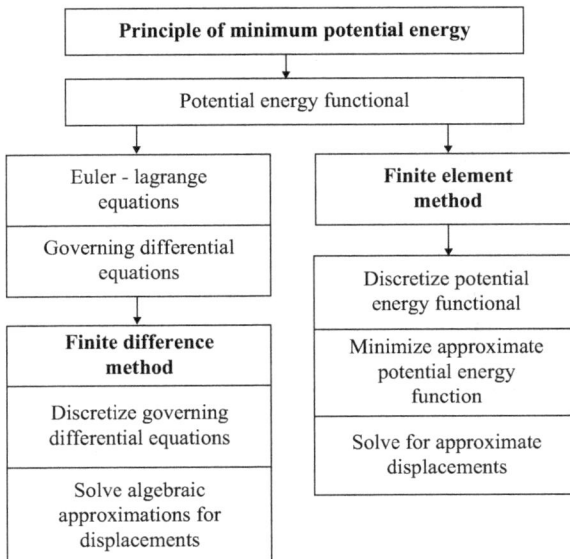

Figure 7.4. A schematic of the finite element and finite difference methods.

The starting point for both methods is the creation of the *functional* that expresses the potential energy for the problem. It is this quantity that must be minimized in order to find the displacements that are the solution to this problem. The functional for the plane stress problem is as follows (Dow 1999):

$$V = \int_{\Omega} \left\{ \frac{1}{2}\left(\frac{E}{1-v^2}\right) \left[\begin{array}{l} \left(u_{,x}^2 + 2vu_{,x}v_{,y} + v_{,y}^2\right) + \left(\dfrac{1-v}{2}\right) \\ \left(u_{,y}^2 + 2vu_{,y}v_{,x} + v_{,x}^2\right) - \left(P_x u + P_y v\right) \end{array} \right] \right\} d\Omega \quad (7.1)$$

Equation 7.1 is classified as a functional because it is *a function of functions*. When Equation 7.1 is examined, we see that it fits this category. This equation depends on the two displacement functions, u and v, and their derivatives over the domain of the problem, which is denoted as Ω. This means that the functions for both u and v that minimize the potential energy functional must be found to solve this problem exactly. In both the finite difference and finite element methods, the functions are approximated by finding the displacements at a discrete number of points.

Before a finite difference model can be created, one must identify the governing differential equations and the boundary conditions for the problem. As shown in Figure 7.4, the governing differential equations are found by applying the Euler–Lagrange equations to the potential energy functional.

Application of the Euler–Lagrange equations to a functional identifies the necessary and sufficient conditions for minimizing the functional. In this case, these conditions comprise the governing differential equations and the boundary conditions for the plane stress problem. This approach for finding these equations is used instead of applying equilibrium in order to show that both the finite difference and finite element methods are solving the same problem, namely, minimizing the potential energy in the problem. The Euler–Lagrange equations are derived and demonstrated in Chapter 2—The Calculus of Variations and in Chapter 3—The Plane Stress Problem of Dow (1999).

The application of the Euler–Lagrange equations to the potential energy functional for the plane stress problem produces the following equilibrium equations in the x and y directions:

$$\left(\frac{E}{1-v^2}\right)\left[u_{,xx} + v\,v_{,xy} + \left(\frac{1-v}{2}\right)\left(u_{,yy} + v_{,xy}\right)\right] = p_x$$

$$\left(\frac{E}{1-v^2}\right)\left[v_{,yy} + v\,u_{,xy} + \left(\frac{1-v}{2}\right)\left(v_{,xx} + u_{,xy}\right)\right] = p_y \quad (7.2)$$

where p_x and p_y are the distributed loads in the x and y directions over the domain of the problem.

Another significant difference between the finite element and finite difference methods can be seen by comparing Equations 7.1 and 7.2. The functional given by Equation 7.1 contains the functions u and v and their first derivatives. The governing differential equations given by Equation 7.2 contain the second derivatives of the functions u and v. These differences show that the two approximation solution techniques follow different paths in order to form their solutions.

As would be expected in the finite difference method, the derivatives contained in the governing differential equations are approximated with difference approximations. The finite difference method for this problem is developed and applied in detail in "Part IV—The Strain Gradient Reformulation of the Finite Difference Method" (Dow 1999). In this treatment of the finite difference method, it is seen that the finite difference method can also solve practically any problem that can be solved by the finite element method. This extension of the method's capabilities and the fact that it is easier to form finite difference models than finite element models may infuse the method with new life in solid mechanics.

In the application of the finite element method, the domain of the problem is subdivided into relatively simple regions as shown in Figure 7.1. The strain energy expression and the work function due to the loads applied to the problem are formed as presented in Chapter 4. The resulting stiffness matrices and loads are assembled as demonstrated in Chapter 2. As noted in Chapter 2, the minimization procedure that is indicated in the flowchart for the finite element method shown in Figure 7.2 is implicitly contained in the element assembly process. The displacements are found by solving the resulting algebraic equations.

As a further reinforcement of the idea that the finite element and finite difference methods use different approaches to achieve the same objective, it can be seen that the structure of the equations solved by the two approaches are significantly different. The finite difference equations consist of an assembly of equations in the form of rows. The finite element equations are formed from the assembly of rectangular blocks of stiffness properties.

7.4 AN ALTERNATE FORMULATION OF A SMOOTHED STRESS REPRESENTATION

This section develops a smoothed stress representation on the boundary of the internal cutout in the Kirsch problem that includes the zero stress boundary conditions. As a result, this smoothed solution is closer to an actual finite difference solution than the smoothed solution formed by averaging the finite element stresses at the nodes.

This development has two objectives. The first is to show that the smoothed solution formed by averaging the finite element stresses will produce an effective error estimator. This is accomplished by demonstrating that the two smoothed representations approach each other as the model is refined. The resulting error estimator is demonstrated in a later section.

The second objective of this section is to present and explain the procedure for introducing the stress boundary conditions into the smoothed solution. This procedure is presented for two reasons. The first is to show a further separation between the finite element and finite difference methods. This, in turn, reinforces the idea that the error estimator is based on the difference between two dissimilar *chains of thought*.

The second reason is to clarify the procedure for introducing the stress boundary conditions into finite difference models. This feature of the finite difference method is rarely, if ever, discussed in detail in most treatments of the finite difference method. The finite difference method is too important to have a key feature that is not clearly understood. In many presentations of the finite difference method, the treatment of the boundary conditions is so ambiguous as to discourage the use of the method.

The inclusion of boundary stresses introduces nodes that are *not on the domain* of the problem being solved. Examples of such nodes are shown in Figure 7.5 as circles on the interior of the cutout in the Kirsch problem. These nodes are called *fictitious nodes* because they are not on

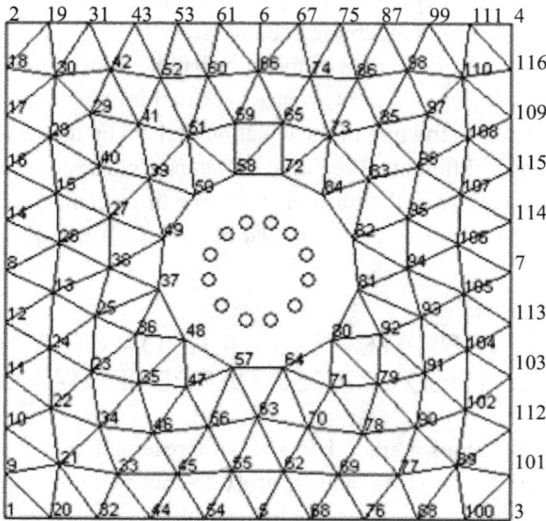

Figure. 7.5. Mesh with fictitious nodes.

the domain of the problem. As can be seen in the figure, one fictitious node is present for each node on the boundary of the circle.

The fictitious nodes are similar to the nodes on the domain of the problem. Each fictitious node has two unknown displacements associated with it. These two unknown displacements provide an avenue for introducing the two boundary stresses into the problem. Since the interior of the circle is a free boundary, the normal stress and the shear stress on the boundary are equal to zero.

In this application, the fictitious nodes are used in conjunction with nodes on the domain of the problem to form a set of nodes that surround each of the nodes on the internal boundary. Since the node of interest is surrounded, it is in the center of the group of nodes. The displacement of this central node and of the nodes surrounding it are used in conjunction with Taylor series expansions to form approximations of derivatives at the central node. Consequently, this group of nodes is referred to as a central difference template. A sample central difference template is shown in Figure 7.6. The central node for this template is the boundary node number 49.

The central difference templates are used in conjunction with the stress transformation given by Equation 6.22 to enforce the normal and shear stresses that exist at each node on the interior of the circle. The three stresses in the x–y coordinate system are expressed in terms of the displacements in the central difference template. The displacements are substituted into the stress transformation given by Equation 6.22 to produce two equations. These equations are used to find the displacements of the fictitious nodes.

Now that we know all of the displacements for the templates associated with a boundary node, we can compute the third stress component at

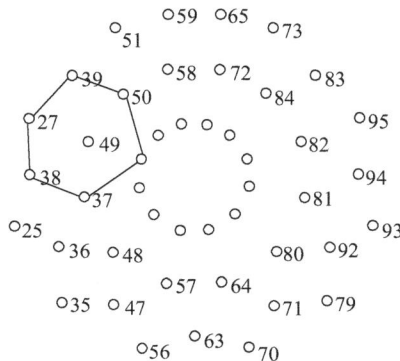

Figure 7.6. Template example.

each node on the circle. The third stress component, the tangential normal stress, is the critical stress component for this problem.

In other words, a smoothed stress representation is formed on the boundary that is a closer representation of a finite difference solution than what would be produced by averaging the finite element stresses. In the next section, the boundary stress found by this procedure will be compared to the stresses formed by smoothing the finite element stresses on the boundary.

7.5   A COMPARISON OF SMOOTHED STRESSES

In the series of figures presented in this section, we will compare the two smoothed stress representations for the stresses on the boundary of the internal cutout in the Kirsch problem as the finite element model is refined. We will see that the smoothed stress formed by averaging the discontinuous finite element nodal stresses become practically indistinguishable from the smoothed stresses found by including the known boundary stresses using the central difference templates as the mesh is refined.

In other words, we are comparing the averaged nodal stresses to nodal stresses that are close to actual finite difference stresses. Since they become practically indistinguishable, we can conclude that the smoothed stress formed by averaging the discontinuous finite element nodal stresses can be considered as an approximate finite difference result. As a result, the smoothed stresses formed from the discontinuous finite element stresses and discontinuous finite element stress representations can be considered as coming from two significantly different solution techniques. Consequently, we can conclude that any differences between the two smoothed solutions are due to errors in the finite element model. The comparison with actual finite difference solutions is left as an exercise for the reader.

The three stress components on the boundary of the internal cutout for the case where the nominal mesh size is h0 = 0.2 are presented in Figure 7.7. The three plots have the same scale so they can be easily compared. As would be expected, both the shear and normal boundary stresses for the finite difference model are exactly zero since these quantities were imposed on the boundary with the central difference templates. As would be expected, the finite element representations are not zero because the boundary conditions are not imposed on finite element models. In fact, these boundary stresses will only reach zero in a fully converged finite element solution.

In the case of the critical stress, namely, the normal stress that is tangent to the surface of the circle, the two approximations have the same

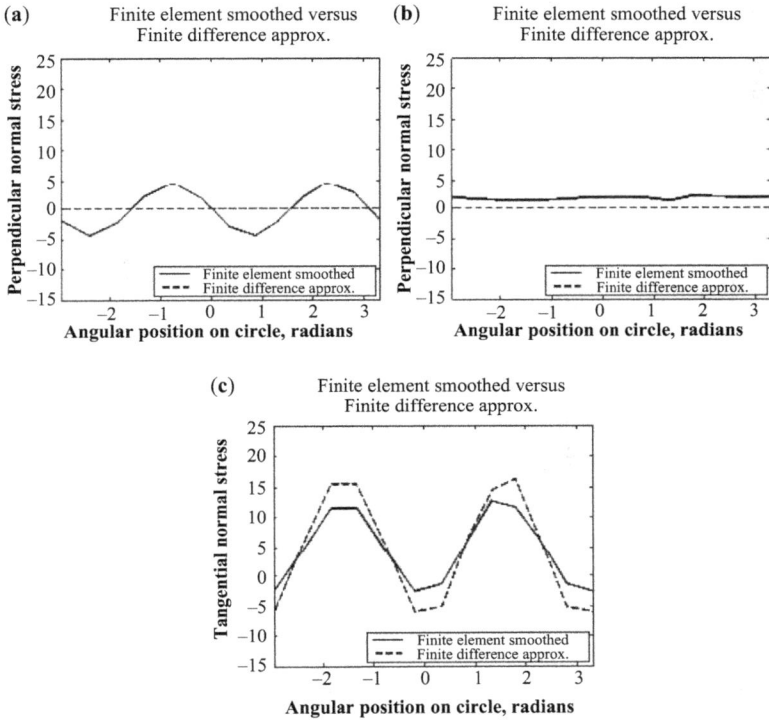

Figure 7.7. Smoothed stress comparisons for h0 = 0.2 units: (a) boundary shear stress, (b) boundary normal stress, and (c) critical normal stress.

shape, but they are not identical. The conclusion to be drawn from this is that the approximate finite difference representation formed using the fictitious points is probably a better representation of the actual finite difference result than is the smoothed solution formed by averaging the finite element stiffness results.

The three stress components on the boundary of the internal cutout for the case where the nominal mesh size is h0 = 0.1 are presented in Figure 7.8. Again, both the shear and normal boundary stresses for the finite difference model are exactly zero since these quantities were imposed on the problem. Again, the finite element representations are not zero. However, they are closer to zero than they were in the previous mesh.

More importantly, the critical stresses for the two smoothed solutions are closer to each other than they were in the previous case. In addition, the maximum critical stress has increased for both smoothed solutions. As the mesh is further refined and the errors are estimated, we will see that

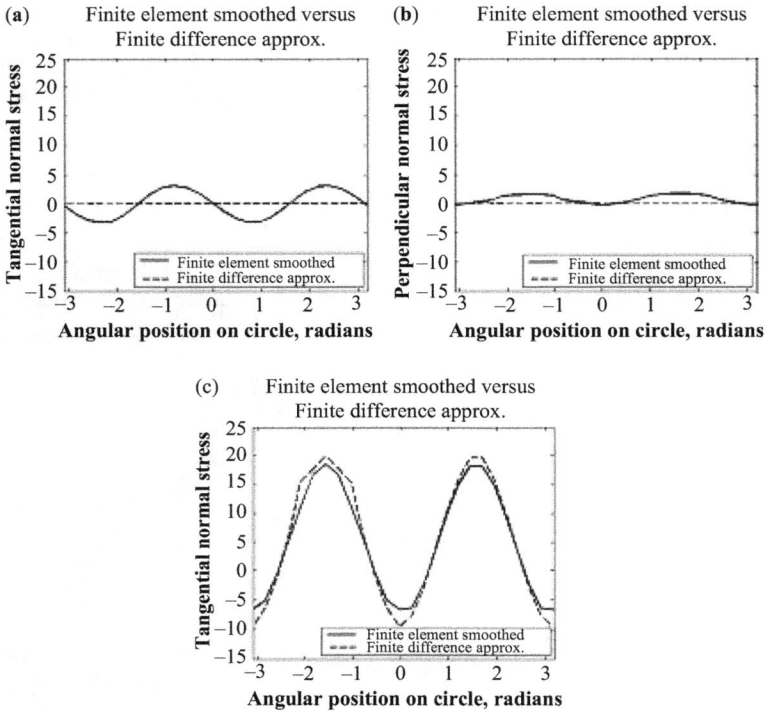

Figure 7.8. Smoothed stress comparisons for h0 = 0.1 units: (a) boundary shear stress, (b) boundary normal stress, and (c) critical normal stress.

the maximum value for the critical stress is closer to the converged value than it was in the previous mesh.

The three stress components on the boundary of the internal cutout for the case where the nominal mesh size is h0 = 0.025 are presented in Figure 7.9. Again, both the shear and normal boundary stresses for the finite difference model are exactly zero since these quantities were specified. As in the previous cases, the finite element representations are not zero. Again, they are closer to zero than they were in the previous mesh. More importantly, the critical stress has gotten larger and the two smoothed solutions are nearly identical.

These results lead to the conclusion that the smoothed solution formed from the finite element stresses leads to acceptable error estimates. These error estimates are used as termination criteria. If the errors are below some predefined level for every element, the analysis is terminated because the solution is deemed accurate enough for use.

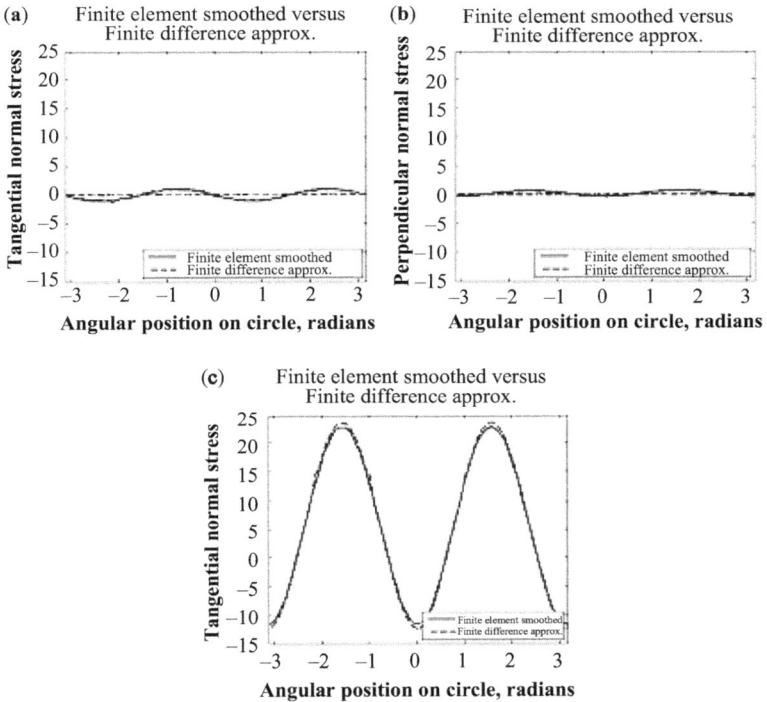

Figure 7.9. Smoothed stress comparisons for h0 = 0.025 units: (a) boundary shear stress, (b) boundary normal stress, and (c) critical normal stress.

7.6 FORMULATION OF AN ERROR ESTIMATOR

In this section, an error estimator for evaluating the modeling effective-ness of individual elements is defined. This error estimator evaluates the error in the ε_x strain component of each element.

This metric is chosen for this presentation because the stresses at the critical point in the Kirsch problem are primarily due to this strain component. As a result, this becomes the simplest metric with which to demonstrate the efficacy of error estimators formed using smoothed finite element results.

One of the most useful characteristics of this approach to forming error estimators is that the errors can be estimated in practically any quan-tity based on stresses or strains. For example, a metrics based on the mag-nitude of the principal stresses in a material or the failure criteria based on shear quantities could be used for ceramic materials.

The error estimates for the individual three-node elements are computed in the following way. The difference between the finite element strain component ε_x and the smoothed strain component ε_x is computed at each node in the element. Then, the largest magnitude of this difference for each element is taken as the error in the element. The error is then normalized with respect to the largest magnitude of ε_x that exists in the model and is expressed as a percentage.

In equation form, the error measure for an individual element in the normal strain in the x direction is the following:

$$\text{Percent Error}\,\varepsilon_x\,(i) = \frac{\max[\text{abs}\,((\varepsilon_x\,(i) - \varepsilon_{x\,\text{ave}}\,(j)))]}{\max[\text{abs}\,\varepsilon_x\,(\text{all elements})]} \times 100 \qquad (7.3)$$

where i refers to the element number and j refers to the smoothed strains at the three element nodes.

The errors at the nodes can be used in constant strain and linear strain elements, that is, three- and six-node triangles, because the largest magnitudes must exist at the nodes of the element. In the case of quadratic strain elements, that is, 10-node triangles, the maximum stresses and strains can exist on the interior of the element. However, this would only be the case at maximum or minimum points.

7.7 APPLICATION OF AN ERROR ESTIMATOR

In this section, the error estimator formed in the previous section is applied to a series of refined meshes for the Kirsch problem. This error estimator evaluates the error in the ε_x strain component of each element.

The first model evaluated is formed with h0 = 0.2 as the basis for the mesh generator that was presented in Chapter 5. The estimated errors are presented in Figure 7.10. Figure 7.10a contains the error estimates for the ε_x component in the normal strain in every element on the mesh. As would be expected, the maximum errors occur in the region of the critical stresses. The maximum estimated errors of 59 and 61 percent are shown at the top and bottom of the internal cutout. Note that the errors on the outer boundary of the square domain range from 0 to 19 percent.

The errors in the elements that form the boundary of the internal cutout are shown in Figure 7.10b. These boundary errors range from 6 to 61 percent. This magnitude of the maximum error makes this result unacceptable for evaluating a design.

(a)   Strain error estimates, ε_x (b)   Strain error estimates, ε_x

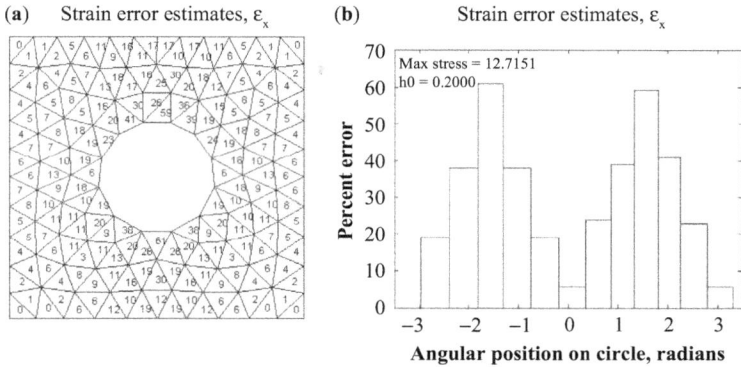

Figure 7.10. Elemental strain errors in σ_x: (a) errors on problem domain and (b) errors on circle boundary.

(a)   Strain error estimates, ε_x (b)   Strain error estimates, ε_x

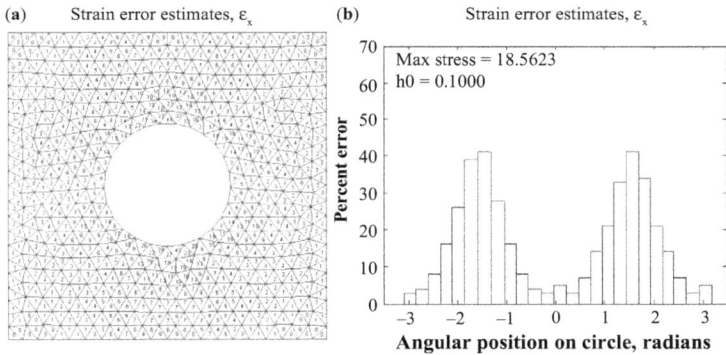

Figure 7.11. Elemental strain errors: (a) errors on problem domain and (b) errors on circle boundary.

When the nominal size of the elements in the mesh is halved by setting h0 = 0.1 in the mesh generator, the estimated errors are presented in Figure 7.11. As can be seen in Figure 7.11a, the maximum error in the critical strains is 41 percent in both the top and bottom locations of the cutout. The errors in the elements that form the boundary of the internal cutout are shown in Figure 7.11b. These boundary errors range from 3 to 41 percent.

The maximum error of 41 percent is a reduction from the maximum of 61 percent that existed in the previous model. This maximum error is still unacceptably high. An acceptable value would depend on many factors such as the size of the safety factor used in the design. Note that the maximum stress has increased to 18.56. In the previous model, the maximum stress was 12.71. In addition, the errors away from the stress concentration and on the outer boundary have been reduced from a maximum of 19 percent to a maximum of 9 percent

The estimated errors are presented in Figure 7.12 for the case where the nominal size of the elements in the mesh is halved again by setting h0 = 0.05 in the mesh generator. Here, the maximum error in the critical strains is 27 percent. This can be seen in the portion of the mesh at the top of the cutout presented in Figure 7.12a. The mesh in this model is so fine that the estimated errors are only discernable if a section of the mesh is expanded. The errors in the elements that form the boundary of the internal cutout are shown in Figure 7.12b. These boundary errors range from 2 to 27 percent.

This is a reduction from the maximum of 41 percent that exists in the previous model. This error of 27 percent is still unacceptably high. Note that the maximum stress has now increased to 21.68. In the previous model, the maximum stress was 18.56. As can be seen, the errors away from the stress concentration continue to shrink.

The sequence of results just presented has shown that the error estimates decrease as the mesh is refined as would be expected. We could continue to make incremental changes in the size of the mesh until an acceptable level of error has been achieved. Instead, we will present the results for a level of mesh refinement that produces an acceptable level of error.

When the nominal size of the elements in the mesh is reduced by a factor of 10 by setting h0 = 0.005 in the mesh generator, the estimated

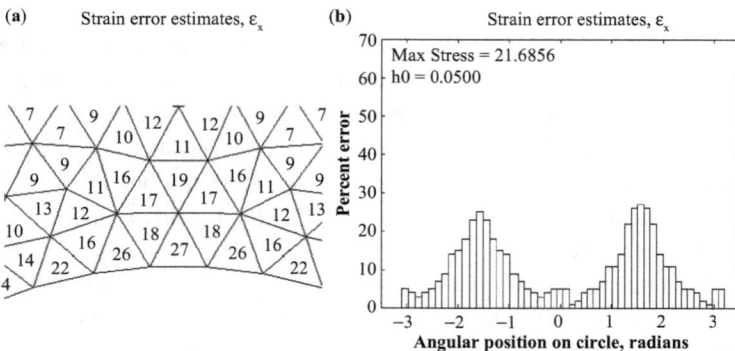

(a) Strain error estimates, ε_x (b) Strain error estimates, ε_x

Max Stress = 21.6856
h0 = 0.0500

Percent error

Angular position on circle, radians

Figure 7.12. Elemental strain errors: (a) errors on problem domain and (b) errors on circle boundary.

errors are presented in Figure 7.13. The maximum error in the critical strains is 4 percent. This can be seen in the portion of the mesh presented in Figure 7.13a. The errors in the elements that form the boundary of the internal cutout are shown in Figure 7.13b. These boundary errors range from 1 to 4 percent. The error in the critical region at the top of the cutout is 2 percent.

This results in a reduction from the maximum of 27 percent that existed in the previous model. An error of 2 to 4 percent is an acceptable value for design purposes. Note that the maximum stress has increased to 24.01. In the previous model, the maximum stress was 21.68.

The upper left-hand corner of the mesh for this model is shown in Figure 7.14. The purpose of this figure is to show that the errors away from the stress concentration featured in Figure 7.13 are small.

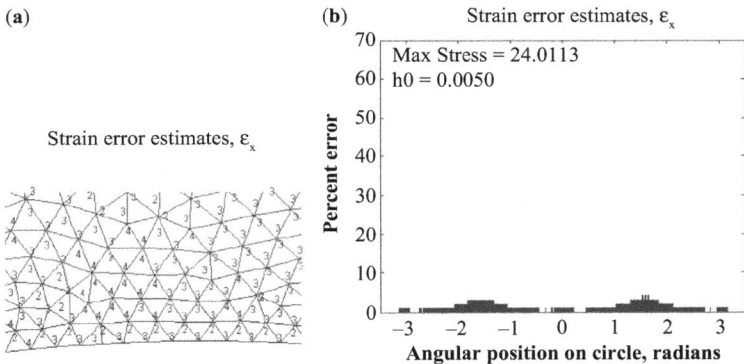

Figure 7.13. Elemental strain errors: (a) errors on problem domain and (b) errors on circle boundary.

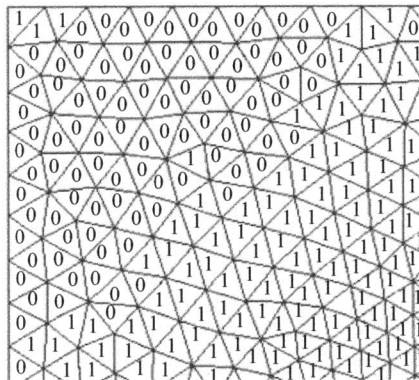

Figure 7.14. Elemental strain errors.

7.8   SUMMARY AND CONCLUSIONS

In Chapter 6, we saw that the interelement jumps in stresses and strains were reduced as the mesh was refined. These jumps are interpreted as identifying when an element cannot capture the complexity of the stresses and strains that exist in the exact solution on its domain.

In this chapter, we demonstrated that these jumps could be exploited to form reliable error estimators with a solid theoretical foundation. This was accomplished by showing that a smoothed solution could be formed from the discontinuous finite element solution and that this could be taken as an approximate finite difference solution.

The difference between the approximate finite difference solution and the discontinuous finite element solution provides an estimate of the error in the finite element solution because both the finite element and the finite difference solutions converge to the exact solution as the mesh is improved. Since the two approximate methods seek their solutions in significantly different ways, the difference between the two solutions indicates flaws in the discrete model.

In the next chapter, the results produced by the error estimator will be used as the basis for identifying elements that must be subdivided in order to achieve a desired level of error in the finite element solution. In addition, the error estimates will be used to identify the level of refinement needed in order to reduce the error to the desired level in only a few refinements.

7.9   EXERCISES

1. Form the error estimators over the full domain of the problem.
2. Form the finite difference version of the smoothed solution with finite difference templates as a set of simultaneous equations where the finite difference templates contain three fictitious nodes.
3. Solve a sequence of refinements of the Kirsch problem with the finite difference method and compare the stress results to the smoothed solution formed by averaging the finite element stress results.

CHAPTER 8

SIMPLE AND EFFECTIVE REFINEMENT GUIDES

8.1 INTRODUCTION

In the previous chapter, a procedure was created for estimating the magnitude of the errors in individual elements. When a model contains elements with high levels of error, these elements must be subdivided in order to improve the representation. In this chapter, a procedure is developed and demonstrated to identify the number of subdivisions that must be given to high-error elements in order to achieve the desired level of accuracy. The insight that provides the basis for the *refinement guide* is seen in Figure 8.1.

Figure 8.1 presents the tangential normal stress on the internal cutout of the Kirsch problem for a sequence of mesh refinements. This rapidly changing stress distribution contains the critical stress for this problem. As can be seen in Figure 8.1, the maximum stress increases in magnitude as the mesh is refined. The increasing maximum stresses in the three figures are 17.21 units, 20.93 units, and 23.98 units, respectively.

The reason for the correlation between the accuracy, which is embodied by the size of the interelement jumps, and the element size was identified in the previous chapter. The interelement jumps decrease as the high-error elements are subdivided. This occurs because the portion of the exact solution that each subdivision represents is smaller than was represented by the original element. As a result, the contribution of the higher-order Taylor series terms in the exact solution that the element must represent are getting smaller since the size of the element is smaller. As a result, the elements are better able to represent the underlying exact solution and the interelement jumps are reduced.

We will use this knowledge in order to create a simple and efficient refinement guide. The refinement guide estimates the number of

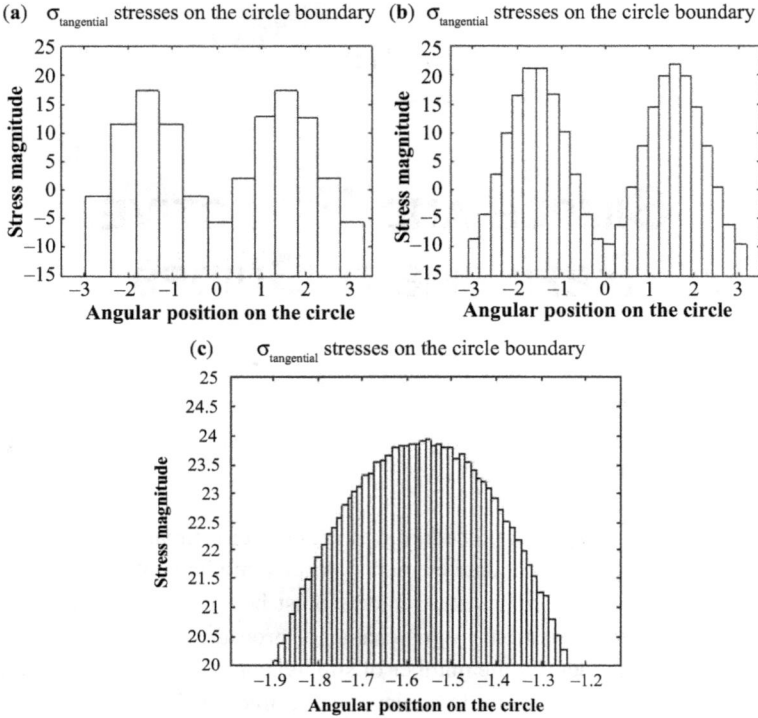

(a) $\sigma_{\text{tangential}}$ stresses on the circle boundary   **(b)** $\sigma_{\text{tangential}}$ stresses on the circle boundary

(c) $\sigma_{\text{tangential}}$ stresses on the circle boundary

Figure 8.1. Tangential normal stress distributions: (a) nominal mesh size, h0 = 0.2; (b) nominal mesh size, h0 = 0.1; and (c) nominal mesh size, h0 = 0.005.

subdivisions that must be given to the high-error elements in order to reduce the level of error in the model to a specified level. The refinement guide is formed by comparing the modeling capability of an element to an approximation of the exact solution that it is trying to represent.

8.2   A REPRESENTATION OF THE MODELING ERROR

We will now form a representation of the approximate modeling error that exists in an individual finite element. This will be done by first forming a Taylor series approximation of the exact solution that exists on the domain of an element. This approximation is then compared to the modeling capability of the element.

The development presented here is demonstrated by applying it to the normal strain in the x direction. This component was chosen because it is the largest component of the critical stress in this problem. Analogous procedures apply to the other strain components as well as to the stress components. Evaluating another variable would only be repetitious.

The exact representation of the normal strain in the x direction that an individual element is attempting to represent can be expressed with the following infinite Taylor series expansion:

$$\varepsilon_x(x,y) = (\varepsilon_x)_0 + (\varepsilon_{x,x})_0 x + (\varepsilon_{x,y})_0 y + \tfrac{1}{2} (\varepsilon_{x,xx})_0 x^2 + (\varepsilon_{x,xy})_0 xy + \tfrac{1}{2} (\varepsilon_{x,yy})_0 y^2 + \tag{8.1}$$

Since a three-node element can only represent the constant strain terms, the errors in the finite element representation of ε_x are due to the inability of the individual elements to model the higher-order terms. When the constant strain component is removed from Equation 8.1, the components that an element cannot represent are left. In other words, the error in a three-node element is produced by the higher-order terms that it cannot represent, which is given as follows:

$$\text{Error}(\varepsilon_x) = (\varepsilon_{x,x})_0 x + (\varepsilon_{x,y})_0 y + \tfrac{1}{2} (\varepsilon_{x,xx})_0 x^2 + (\varepsilon_{x,xy})_0 xy + \tfrac{1}{2} (\varepsilon_{x,yy})_0 y^2 + \tag{8.2}$$

In the refinement guides developed here, we will assume that the majority of the error is due to the inability of an element to represent the lowest-order terms in the error expressions. That is to say, the majority of the error in a *three-node element* is assumed to exist because the element cannot represent the linear terms in the strain representations. For the strain component under consideration here, the assumed error is given as follows:

$$\text{Assumed Error}(\varepsilon_x) = (\varepsilon_{x,x})_0 x + (\varepsilon_{x,y})_0 y \tag{8.3}$$

The significant feature of Equation 8.3 is the fact that the errors vary linearly with the dimensions of the element being analyzed. As the element is subdivided, the dimensions of the element are, by definition, being reduced. Consequently, the errors are reduced. Note that the contribution of the quadratic and higher-order terms will decrease even faster than the linear terms as the element size is reduced. The errors in the ε_y and γ_{xy} strain components and the three stress components for a three-node element have a similar linear form.

In contrast, if six-node elements are used to form a finite element model, these elements can represent the linear strain terms in addition to the constant strain terms. As a consequence, the assumed error in a six-node element is due to the quadratic terms that the element cannot represent. That is to say, the errors in a *six-node element* representation of

$\varepsilon_x(x,y)$ are assumed to be largely due to the element's inability to represent the following quadratic terms:

$$\text{Assumed Error} (\varepsilon_x) = \tfrac{1}{2} (\varepsilon_{x,xx})_0 \, x^2 + (\varepsilon_{x,xy})_0 \, xy + \tfrac{1}{2} (\varepsilon_{x,yy})_0 \, y^2 \quad (8.4)$$

In this case, the error terms vary quadratically with the dimensions of the element being evaluated. Since the errors in six-node elements decrease quadratically instead of linearly, a refinement of a six-node element produces faster improvements in the solution than does the refinement of a three-node element.

In brief, the refinement guide developed here relates the modeling capability of an individual element to the assumed error in the portion of the exact solution that underlies the element. After the assumed error is computed, the number of subdivisions needed to reduce the interelement jumps between elements to below the prescribed level of acceptable error is estimated.

8.3   A LIMITATION OF THE REFINEMENT GUIDE

The refinement guides developed and demonstrated in later sections have a built-in limitation. An example of the source of this deficiency can be seen in Figure 8.2. In this figure, a bold line outlines a group of elements that surround an element that contains an unacceptable level of error. The element that is to be subdivided is designated as element 222. This group of elements is contained in a uniform mesh with h0 specified in the mesh generator as 0.01 units.

In this example, the magnitudes of the higher-order strain gradient terms that are assumed to be the major source of errors in element 222 are derived from the group of elements that surround the element. Since these terms are computed from a region that is larger than the element itself, the computation is not focused directly on the high-error element. As a result, the terms assumed to produce a majority of the error in an element are not as accurate as would be desired.

In photographic terms, this deficiency can be described in terms of the resolution of an image. In other words, the *pixel* containing the 13 elements in Figure 8.2 that is used to compute the estimate of the underlying exact solution is larger than the element being evaluated. Consequently, the desired result is somewhat *blurred*. As we will see in a later section, the resolution of the error-producing terms in the underlying exact solution sharpens as the mesh is refined and the size of the *pixel* gets smaller.

The source of the lack of focus on the element being subdivided can be seen in the following computation. The estimates of the Taylor series

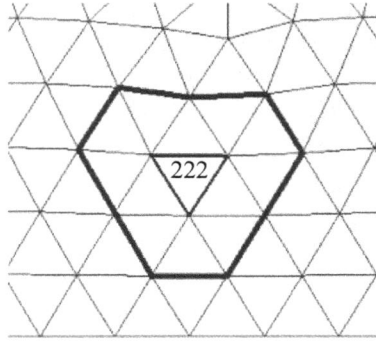

Figure 8.2. Element patch.

coefficients of the *exact solution that underlies* the element being sub-divided are computed with the following equation:

$$\{\varepsilon,\} = [\Phi]^{-1} \{d\} \tag{8.5}$$

where $\{\varepsilon,\}$ is the vector of the estimated strain gradient quantities that the element is trying to represent, $\{d\}$ is the vector of nodal displacements for the group of elements, and $[\Phi]$ is the matrix of nodal locations for the patch of elements.

Equation 8.5 is identical in form to Equation 4.12, which is used to generate an element stiffness matrix. For the case shown in Figure 8.2, this equation is based on the nodal locations for 13 elements. As a result, the Taylor series coefficients apply to the patch of elements and not just to the element being evaluated.

In addition to identifying a limitation in the refinement guide, the development presented in this section provides the basis for another approach for forming error estimates. The computation contained in Equation 8.5 provides an estimate of the coefficients contained in Equation 8.3. These coefficients are not used directly in the error estimator developed here. However, the investigation of their use in an error estimator is a possible research topic.

8.5 A SIMPLER COMPUTATION OF THE ASSUMED STRAIN ERROR TERMS

In this section, simpler and more efficient procedure is presented for computing the coefficients of Equation 8.3. These coefficients are the strain gradient terms that are assumed to produce a majority of the errors in individual three-node finite elements. This computation differs significantly from the procedure presented in the previous section. In this computation, the error-producing terms are based on the smoothed strains at the three nodes of the element being evaluated. As a result, the matrix that must be inverted is only a three-by-three matrix. However, the computation must be performed for each of the three strain components.

In the previous section, the error-producing terms are based on the nodal displacements in the patch of elements surrounding the element being evaluated. This means that one larger matrix must be inverted in order to get the desired error-producing terms.

The procedure presented in this section possesses the same limitation as the procedure presented in the previous section that uses the patch of elements. The approximation of the exact solution is not fully focused on the element being evaluated. This is the case because the smoothed strains used to compute the error-producing terms are themselves computed from the nodal displacements contained in a patch of elements that surround the element being evaluated.

The approximations of the strain gradient terms that are assumed to produce a majority of the errors in a finite element result are developed for three-node elements and are demonstrated for the Kirsch-like problem. The approximations are extracted from the Taylor series representation of the averaged nodal strains for the element. The approximation of each strain component contains three terms because the averaged strains are available at the three nodes of the finite element.

The Taylor series expansion for the strain component in the x direction, ε_x, formed from to these averaged strains is the following:

$$\varepsilon_x(x,y) = (\varepsilon_x)_0 + (\varepsilon_{x,x})_0 \, x + (\varepsilon_{x,y})_0 \, y \qquad (8.6)$$

where $\varepsilon_x(x,y)$ is the strain at the point (x,y), $(\varepsilon_x)_0$ is the strain gradient quantity that designates the constant strain state, $(\varepsilon_{x,x})_0$ is the strain gradient quantity that designates the rate of change of ε_x in the x direction, and $(\varepsilon_{x,y})_0$ is the strain gradient quantity that designates the rate of change of ε_x in the y direction.

When Equation 8.6 is used to create the transformation for computing the magnitudes of the strain gradient components that are due to the average strains at the nodes, we have the following:

$$\begin{Bmatrix} (\varepsilon_x)_1 \\ (\varepsilon_x)_2 \\ (\varepsilon_x)_3 \end{Bmatrix} = \begin{bmatrix} 1 & x_1 & y_1 \\ 1 & x_2 & y_2 \\ 1 & x_3 & y_3 \end{bmatrix} \begin{Bmatrix} (\varepsilon_x)_0 \\ (\varepsilon_{x,x})_0 \\ (\varepsilon_{x,y})_0 \end{Bmatrix} \qquad (8.7)$$

where $(\varepsilon_x)_i$ = The averaged nodal strain at node i of the element; x_i and y_i are the local coordinates of node i; and $(\varepsilon_x)_0$, $(\varepsilon_{x,x})_0$, and $(\varepsilon_{x,y})_0$ are the strain gradient quantities at the origin of the element.

The strain gradient quantities are found by inverting Equation 8.7. The result of this operation is the following:

$$
\begin{Bmatrix} (\varepsilon_x)_0 \\ (\varepsilon_{x,x})_0 \\ (\varepsilon_{x,y})_0 \end{Bmatrix} = \begin{bmatrix} 1 & x_1 & y_1 \\ 1 & x_2 & y_2 \\ 1 & x_3 & y_3 \end{bmatrix}^{-1} \begin{Bmatrix} (\varepsilon_x)_1 \\ (\varepsilon_x)_2 \\ (\varepsilon_x)_3 \end{Bmatrix}
\tag{8.8}
$$

The strain gradient quantities computed by Equation 8.8 are the Taylor series coefficients that represent the ε_x strain component in the approximation of the exact solution against which the element being evaluated is compared. The two higher-order terms, $(\varepsilon_{x,x})_0$ and $(\varepsilon_{x,y})_0$, approximate complexities in the exact solution that an individual three-node element cannot represent. These terms are assumed to be the source of the majority of the errors in the finite element strain representations.

8.6 BEHAVIOR OF THE CRITICAL STRAIN ERROR TERMS

This section presents the characteristics of the estimates of the higher-order terms that are assumed to cause a majority of the errors in three-node elements. This presentation shows that the resolution of these high-order terms improves as the mesh is refined. This implies that the refinement guides more accurately predict the number of subdivisions needed to produce converged finite element results as the mesh is refined. This improvement occurs because the patch of elements used to compute these quantities is getting smaller.

As discussed earlier, the inability of three-node constant strain elements to represent linear strain variations is assumed to make the largest contribution to the errors in these elements. The objective of this section is to explain and demonstrate the behavior of these linear strain variations.

As we will see, two characteristics of the linear strain gradient terms might seem counterintuitive. On the one hand, these terms can be close to or equal to zero in regions of high error. This occurs if either a maximum or minimum point is located on the element. This is the case because the test for a maximum or a minimum is for the first derivative to equal to zero.

On the other hand, these linear terms often increase as the mesh is refined. This increase occurs if an element that is subdivided envelops a region of rapidly changing strain. The initial element contains a weighted average of the rate of change on its domain. As an element is subdivided, one or more of the subelements can represent regions where the change is greater than the average. As a result, the linear strain gradient term will increase in these elements. However, the overall contribution to the elemental error will be smaller because the size of the element has been reduced.

This presentation focuses on the behavior of the $(\varepsilon_{x,x})_0$ and $(\varepsilon_{x,y})_0$ terms. The strain component ε_x is featured because it makes the largest contribution to the stress concentration that exists in this problem. A discussion of the critical error terms for the other strain components would not add significant content. Figures 8.3 to 8.5 shows the values of the linear variations in ε_x, $(\varepsilon_{x,x})_0$ and $(\varepsilon_{x,y})_0$ that are contained in the elements in the region of the stress concentration on the lower edge of the cutout in the Kirsch problem.

Plots of the strain gradient terms $\varepsilon_{x,x}$ and $\varepsilon_{x,y}$ on the bottom edge of the cutout for the finite element model with a nominal mesh size of $h0 = 0.20$ are shown in Figure 8.3. The location of the stress concentration is marked by the x in these figures. The element containing the stress concentration is outlined. We will focus on the behavior of the linear strain gradient terms in the region of the element containing the stress concentration.

Of special interest is the fact that the $\varepsilon_{x,x}$ term shown in Figure 8.3a for the element containing the stress concentration has a value of zero in the element. As shown in Figure 7.10, the error estimate for ε_x in this element

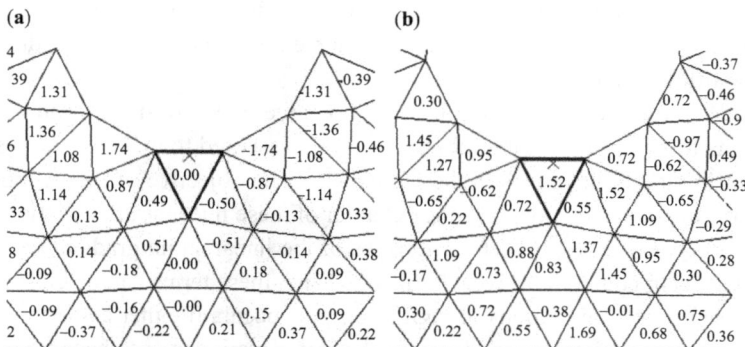

Figure 8.3. Second-order strain gradient terms, $h0 = 0.2$: (a) strain gradient term, $\varepsilon_{x,x}$ and (b) strain gradient term, $\varepsilon_{x,y}$.

is 61 percent. It may seem counterintuitive that a zero or a low value exists for the term that is assumed to produce most of the error in an element with a high level of error. The reason for this counterintuitive behavior is easily explained as follows.

As can be seen in Figure 8.3a, the two elements flanking the element containing the stress concentration have values for $\varepsilon_{x,x}$ of $+1.74$ and -1.74, respectively. The fact that these terms change sign as x increases indicates that a maximum value of ε_x exists in the element that is located between these two elements. In other words, the standard test from the Calculus for identifying a maximum or minimum point has identified the location of a local maximum value that we know exists.

The effect of this behavior does not debilitate the adaptive refinement process. At most, this counterintuitive behavior might add an iteration or two until an acceptable result is attained. The need for extra iterations may occur because the estimated error in the element might be somewhat reduced. As a result, fewer subdivisions than would be ideal could be given to the element.

The mesh shown in Figure 8.3 is refined in Figure 8.4 by reducing the nominal mesh size from h0 = 0.2 to h0 = 0.1. Plots of the $\varepsilon_{x,x}$ and $\varepsilon_{x,y}$ strain gradient terms for the refined model are presented in Figures 8.4a and 8.4b, respectively.

In this case, the element containing the stress concentration in the previous mesh has been essentially replaced by four elements. These four elements are outlined with the heavy line in this figure. These four elements do not cover the exact domain of the single element that was outlined in the previous figure, but they are close to covering the original area.

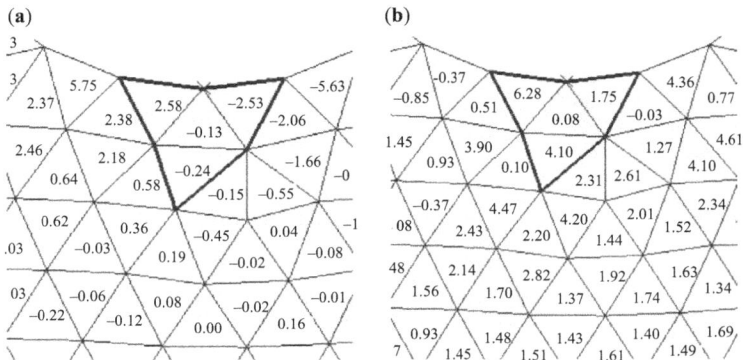

Figure 8.4. Second-order strain gradient terms, h0 = 0.1: (a) strain gradient term, $\varepsilon_{x,x}$ and (b) strain gradient term, $\varepsilon_{x,y}$.

In Figure 8.4a, the two elements on the *boundary of the cutout* with magnitudes of approximately 2.5 represent the portion of the boundary covered by the initial element. These two elements contain the other characteristic of the linear strain gradient terms that might seem counterintuitive. That is to say, the magnitudes of these terms are larger than they were in the element they replaced.

The fact that these two elements display opposite signs indicates they are representing the region containing a maximum value of ε_x. In contrast to Figure 8.3, the critical point is located at a nodal point instead of on the boundary of an element.

The increase in the linear strain gradient term as the mesh is refined also exists in the other two elements shown on the boundary of the cutout. The values of these quantities in these elements are $+5.75$ and -5.63, respectively. These elements comprise subdivisions of elements with linear strain variations of $+1.74$ and -1.74, respectively, in the previous mesh. This means that the rate of change in ε_x is higher in this smaller region than it is in the domain of the larger element.

When the behavior of $\varepsilon_{x,y}$ shown in Figure 8.4b is examined, we see that it is similar to that of $\varepsilon_{x,x}$. The magnitudes on the boundary have increased, and one of the elements on the interior of the subdivision has a low value for the term.

When the mesh is refined again by halving the nominal size to h0 = 0.05, the results for the critical error terms are presented in Figure 8.5. The behavior of the linear strain gradient terms mirror the behavior of these terms in the previous two refinements. This can be seen by examining these terms in the element that now contains the stress concentration.

(a)

(b)

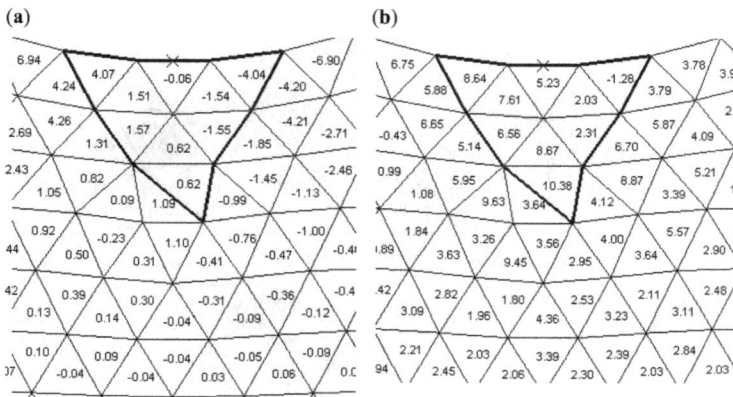

Figure 8.5. Second-order strain gradient terms, h0 = 0.05: (a) strain gradient term, $\varepsilon_{x,x}$ and (b) strain gradient term, $\varepsilon_{x,y}$.

The value of $\varepsilon_{x,x}$ is small because the element contains the stress concentration. The value of $\varepsilon_{x,y}$ is significantly larger than it was in the initial mesh. This is the case because as can be seen in Figure 6.14b, the strain is rapidly increasing in the y direction. In other words, the behavior of the linear strain gradient terms is as expected.

In the next section, we will develop the refinement guides in terms of these higher-order strain gradient terms. Then we will see that the increase in the magnitudes of these terms does not mean that more refinements are necessary. This is the case because the errors produced by the inability of these elements to represent these higher-order terms also depend on the size of the element. Since the size of the element is being reduced, the contribution to the error is reduced. Consequently, the level of refinement is reduced.

8.7 DERIVATION OF A SIMPLE AND EFFECTIVE REFINEMENT GUIDE

We will now derive a simple and effective refinement guide for finite element models formed with three-node constant strain elements. In order to reduce the computation required to form the refinement guide, the level of refinement is found for one component of the stress or strain at a time. This development features the strain component ε_x. When the development of a refinement guide for a three-node element is completed, we will show that the basic idea can be extended to higher-order elements.

As before, we will assume that the majority of the error in ε_x in a three-node element is due to its inability to represent the linear terms in the Taylor series expansion of the strains, namely, $(\varepsilon_{x,x})_0$ and $(\varepsilon_{x,y})_0$. As a result, the estimated error in the element can be taken to be as follows:

$$E_{estimated} = d_{elemental} \left((\varepsilon_{x,x})_0 + (\varepsilon_{x,y})_0 \right) \tag{8.9}$$

where $d_{elemental}$ = the distance to the furthest point in the element from the local origin.

In Equation 8.9, the x and y distances from the local origin of the element to the most distant point are taken to be the same. This assumes that the element approximates an equilateral triangle, which, in turn, is the triangular equivalent of a circle. This concept was discussed in Chapter 5, where the quality of triangular geometry was quantified.

A refinement guide that is developed in Dow (1999) does not make the assumption that the x and y distances to the point furthest from the local origin are the same. This more specialized refinement guide uses the higher-order strain gradient quantities similar to those computed in Equation 8.8 as well as the actual element geometry to estimate the number of needed subdivisions.

The goal of a refinement guide is to subdivide an element so that the subelements contain no more error than the acceptable level. In other words, we need to know the maximum size that an element can have so the error does not exceed this level. We assume that the subdivided elements have the same magnitudes of $(\varepsilon_{x,x})_0$ and $(\varepsilon_{x,y})_0$ as has the element to be subdivided. Therefore, the acceptable error in the subdivided element is related to the sources of the error $(\varepsilon_{x,x})_0$ and $(\varepsilon_{x,y})_0$ as follows:

$$E_{acceptable} = d_{acceptable}((\varepsilon_{x,x})_0 + (\varepsilon_{x,y})_0) \qquad (8.10)$$

where $E_{acceptable}$ = the level of error that is acceptable in the solution.

If we relate the actual size of the element to the acceptable size of the element as follows, we have:

$$D_{elemental} = n\, d_{acceptable} \qquad (8.11)$$

where $(n - 1)$ is the estimated number that the nominal size of element must be subdivided into in order to reduce the error in the element to the specified level of error.

When we divide Equation 8.9 by Equation 8.10 and substitute Equation 8.11, we have the following:

$$\frac{E_{estimated}}{E_{acceptable}} = \frac{n\, d_{acceptable}(\varepsilon_{x,x} + \varepsilon_{x,y})_0}{d_{acceptable}(\varepsilon_{x,x} + \varepsilon_{x,y})_0} = n \qquad (8.12)$$

The number n is used to estimate the maximum distance from the local origin to the most distant point in the element being analyzed. That is to say, the maximum length in the new element that is estimated to reduce the level of error to the acceptable level is given by Equation 8.11 as $d_{acceptable} = d_{actual}/n$.

By assuming that the subdivided elements will have approximately the same shape as the initial elements, the estimated nominal element size h0 in the refined model that will produce a finite element model with an acceptable level of error is $(h0)_{new} = (h0)_{old}/n$.

A refinement guide of the same form applies to the other strain components as well as to the three stress components. This refinement guide will be demonstrated in the next section for both the strain and the stress components in the x direction for the Kirsch problem formed with three-node elements.

Before demonstrating the refinement guide that was just formed, the extension to a six-node element will be presented. In the case of a six-node element, the first set of higher-order strain gradient terms that the element cannot represent is the set of quadratic terms. These terms are the following for the strain component in the x direction: $(\varepsilon_{x,xx})_0$, $(\varepsilon_{x,xy})_0$, and $(\varepsilon_{x,yy})_0$. These quadratic terms are multiplied by x^2, xy, and y^2, respectively. By once again considering that the triangular element is well-conditioned, we can assume that the magnitudes of x and y are similar. When this is the case, the refinement guide for six-node, linear strain elements is the following:

$$\frac{E_{estimated}}{E_{acceptable}} = \frac{n^2 \, h^2_{acceptable}(\varepsilon_{x,xx} + \varepsilon_{x,xy} + \varepsilon_{x,yy})_0}{h^2_{acceptable}(\varepsilon_{x,xx} + \varepsilon_{x,xy} + \varepsilon_{x,yy})_0} = n^2 \qquad (8.13)$$

Equation 8.13 shows that the errors in a six-node linear strain triangle decrease with the square of the size reduction. This means that a model formed from six-node elements requires fewer elements to represent an acceptable result than does a model formed from three-node elements.

A refinement guide based on the direct use of the quadratic strain gradient terms has already been developed and demonstrated for a three-node, linear strain bar element in Dow (1999). However, this type of refinement guide has not yet been demonstrated for a six-node element.

8.8 DEMONSTRATIONS OF A SIMPLE AND EFFECTIVE REFINEMENT GUIDE

In this section, the effectiveness of this simple error estimator is demonstrated for two levels of acceptable error, 10 and 5 percent. These two error levels have not been chosen at random. It is often the case that a finite element result converges at or between these levels of acceptable error. A model is said to have converged if the result of the finite element analysis changes very little between adaptive refinement cycles. Consequently, the result is considered to be acceptable.

If, on the other hand, there is a significant change in the quantity being evaluated between the refinement cycles, an additional refinement cycle is recommended even though the acceptable error criterion has been satisfied. The easiest way to force an additional adaptive refinement cycle is to reduce the level of acceptable error and reapply the adaptive refinement procedure to the current finite element model. If the result of the critical variable does not change significantly, the result can be said to have converged.

In the final analysis, the convergence of the solution overrides the achievement of the acceptable error that has been specified in the adaptive refinement procedure as the accuracy criterion. In other words, *the primary function of the refinement guide* is to reduce the number of iterations needed to approach an acceptable answer. The efficacy of this simple refinement guide will now be demonstrated with three examples.

8.8.1 DEMONSTRATION 1

When the modified Kirsch problem is solved for a nominal mesh of h0 = 0.20 and the acceptable error is given as 10 percent, the results of the first iteration of the adaptive refinement process are presented in Figure 8.6. The estimated error in each of the elements in the finite element model is presented in Figure 8.6a.

As can be seen in Figure 8.6a, the majority of the elements have estimated errors that are equal to or below 10 percent. In other words, they satisfy the specified acceptable error criterion. This is reflected in Figure 8.6b where the elements that satisfy the accuracy criterion have a recommended level of refinement of 1, which means that they do not need to be subdivided.

As expected, the highest levels of errors are in the regions of the stress concentrations that are indicated by the x's on the top and bottom of the circular cutout. This is reflected in Figure 8.6b where the refinement guide indicates that all of the elements that are identified as needing refinement are near the two stress concentrations. The maximum error occurs on the

(a) (b)

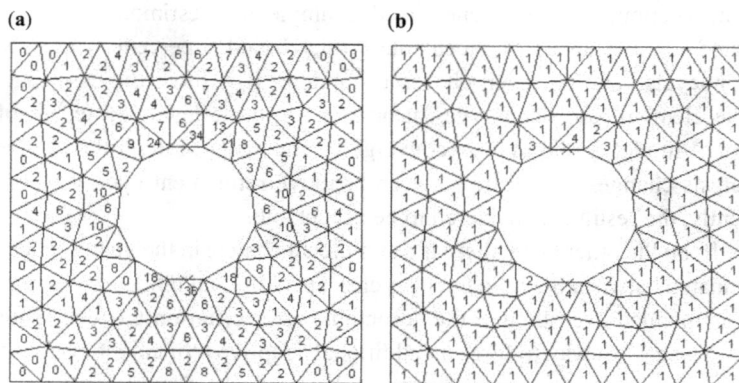

Figure 8.6. Error estimates and refinement guides for σ_x, nominal mesh size = 0.20, acceptable error = 10%: (a) error estimate and (b) refinement guide.

bottom of the cutout and is estimated at 36 percent. The maximum stress in this element is 17.380 units.

The refinement guide recommends that four subdivisions must be given to the two elements with the highest level of estimated error in order to achieve an estimated error of less than or equal to 10 percent. Since the nominal mesh size for this model is h0 = 0.020, the nominal element size that is estimated to reduce the error to the acceptable level is h0 = 0.20/4 = 0.05.

Note that uniform refinement will be used in most of the examples presented here. This is contrary to the primary reason for implementing the adaptive refinement of finite element models, namely, to produce efficient and accurate finite element models. If elements are subdivided that already represent the exact solution with a satisfactory level of accuracy, the subsequent uniformly refined finite element models will become larger than necessary and, hence, inefficient.

Uniform refinement is used in these demonstrations because the primary goal of this book is to present a compact overview of all aspects of the finite element method and adaptive refinement. In order to incorporate a form of adaptive refinement that only refines the elements that exceed the specified level of error, a significantly more complex mesh generator than the one presented in Chapter 5 would have to be introduced. However, the use of uniform refinement in these examples does not detract from demonstrating the effectiveness of the error estimator and refinement guide that are presented here.

When the modified Kirsch problem is solved for the recommended nominal mesh size of h0 = 0.05 and the acceptable error is maintained at 10 percent, the results of the error analysis and the refinement guide are presented in Figure 8.7. Only the lower portion of the mesh is shown in

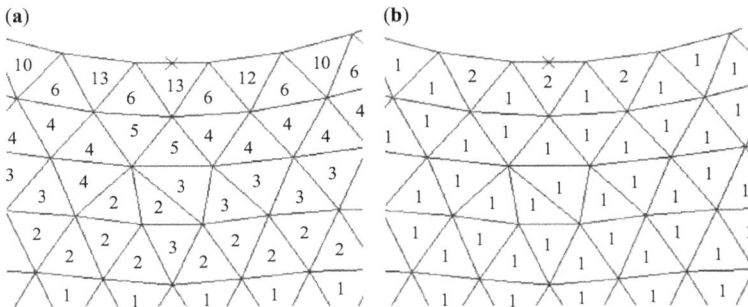

Figure 8.7. Error estimates and refinement guides for σ_x, nominal mesh size = 0.050, acceptable error = 10%: (a) error estimate and (b) refinement guide.

these figures because it contains the largest estimated error. This smaller portion of the mesh is shown because the error estimates and the refinement guides in the figures would be too small to read if the full mesh were presented.

As can be seen in Figure 8.7a, the model has been significantly improved by the reduction in the size of the elements in the initial mesh. The estimate of the maximum error in the element containing the stress concentration has been reduced from 36 to 13 percent. The maximum estimated error is only above the acceptable limit by 3 percent instead of by the 26 percent that occurred in the previous mesh.

The refinement guide estimates that the acceptable error will be achieved by subdividing the element with the maximum error by a factor of two. This contrasts with the previous estimate of four subdivisions. Furthermore, the maximum stress for this model is found to be 23.924 units. This compares to the initial result in the previous case of 17.380 units. This maximum stress result is 27 percent higher than the stress from the previous case.

When the nominal mesh size for the model of the Kirsch problem is reduced to h0 = 0.05/2 = 0.025 as recommended by the refinement guide, the results are presented in Figure 8.8. As we can see in Figure 8.8a, the estimated error in every element is below the acceptable limit of 10 percent. As would be expected, the refinement guide shown in Figure 8.8b does not indicate that any refinement is required to achieve the desired level of estimated error.

In this case, the maximum stress is found to be equal to 24.121 units. When this value is compared to the previous value for the maximum stress of 23.924, the change is equal to 0.8 percent. Consequently, as we

(a)   (b)

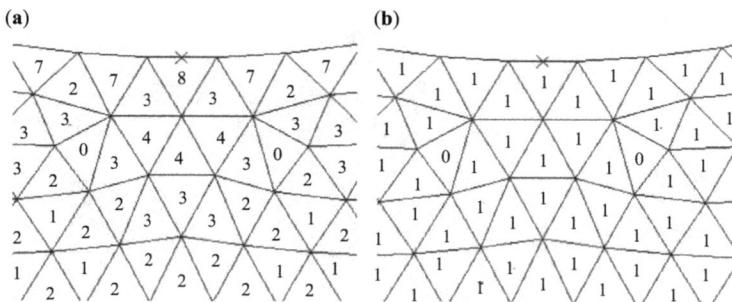

Figure 8.8. Error estimates and refinement guides for σ_x, nominal mesh size = 0.025, acceptable error = 10%: (a) error estimate and (b) refinement guide.

can infer by the small amount of change, the finite element model has converged.

8.8.2 DEMONSTRATION 2

Earlier in the chapter, we observed that the estimated error would be better approximated for finer meshes. This is the case because the error in an element is estimated using the strains in the patch of elements that surround it, as shown in Figure 8.2. This means that the portion of the actual solution being modeled with a finer mesh is smaller than it would be for a coarser mesh. Consequently, the portion of the exact solution being represented by the smaller patch of elements is less complex than the portion of the exact solution being represented by a larger patch. As a result, the low-order modeling capability of an individual element is better able to capture the actual solution. Therefore, the estimated error would be smaller and more accurate.

This concept was validated by the results of the previous example. The mesh formed with the nominal element size of h0 = 0.20 required two applications of the adaptive refinement procedure to produce a model that satisfied the criterion for an acceptable error. However, the mesh formed with the nominal element size of h0 = 0.050 required only one iteration of the adaptive refinement process to achieve an acceptable result.

As a further demonstration of the idea that a smaller initial element size produces more accurate error estimates and, hence, better refinement guides, the Kirsch problem is solved with an initial mesh size of h0 = 0.1 and an acceptable error of 10 percent. This starting point is chosen because the initial mesh size is between h0 = 0.2 and h0 = 0.05. The error estimates and refinement guides are presented in Figure 8.9.

The initial error estimate in the region near the critical stress concentration point exceeds the specified limit of 10 percent. As expected, this maximum estimate of 22 percent is bracketed by the maximum error estimates of the cases with the initial mesh sizes of h0 = 0.20 and h0 = 0.05, which are 36 percent and 13 percent, respectively.

Similarly, the level of refinement identified by the refinement guide is also bracketed by the estimates for h0 = 0.20 and h0 = 0.50. In this case, the recommended number of subdivisions is three instead of four and two, which was recommended in the two previous cases. Furthermore, the critical stress found for this model is 21.111 units. This stress value is higher than the initial value of 17.380 units that was found for the model with h0 = 0.20. It is also lower than the value for the nominal element size of 0.050 of 23.924 units.

(a)   (b)

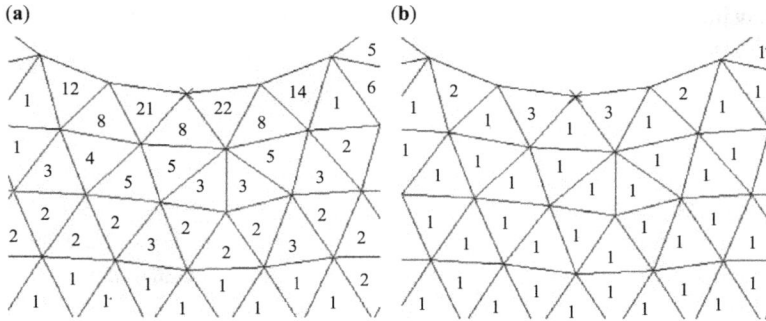

Figure 8.9. Error estimates and refinement guides for σ_x, nominal mesh size = 0.10, acceptable error = 10%: (a) error estimate and (b) refinement guide.

(a)   (b)

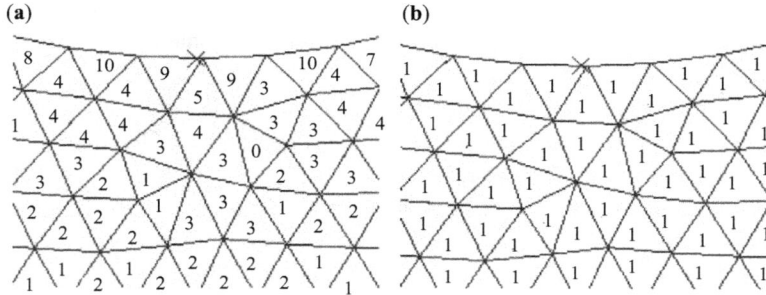

Figure 8.10. Error estimates and refinement guides for σ_x, nominal mesh size = 0.0333, acceptable error = 10%: (a) error estimate and (b) refinement guide.

When the mesh refinement value specified by the refinement guide of three subdivisions is applied, the mesh size that is estimated to produce the acceptable level of error of 10 percent or less is equal to h0 = 0.10/3 or h0 = 0.0333. When the finite element model with this mesh size is formed and solved, the resulting the error estimates and the refinement guides are presented in Figure 8.10.

As we can see in Figure 8.10a, the estimated error in every element is equal to or below the acceptable limit of 10 percent. As would be expected, the refinement guide shown in Figure 8.10b does not indicate that any refinement is required.

In this case, the maximum stress is found to be equal to 23.698 units. When this value is compared to the previous value for the maximum stress of 21.111 units, the change is equal to 10.1 percent. Although this model satisfied the criterion specified by the specified acceptable error of 10 percent, such an amount of change might recommend that another refinement be applied.

The results for such a refinement are available in Figure 8.8. In this case, the nominal mesh size is h0 = 0.25 and the maximum stress is 24.121 units. This represents a change of 1.8 percent from the 23.698 that was found when h0 = 0.0333. This result shows that the maximum stress value of 23.698 found for the mesh shown in Figure 8.10 is a satisfactory result.

8.8.3 DEMONSTRATION 3

We will now look at the behavior of the adaptive refinement process when the acceptable error criterion is tightened from 10 to 5 percent. When a model with a nominal mesh size of h0 = 0.10 with a 5-percent level of acceptable error is solved, the results are presented in Figure 8.11.

The error estimates found for this case are identical to those found for the model shown in Figure 8.9. This is as expected because the nominal element size is identical. However, the results of the refinement guide are different because the acceptable error in this case is specified as 5 percent instead of 10 percent. Consequently, more subdivisions are recommended in order to achieve the desired level of error of 5 percent. In this case, five subdivisions of the nominal element size are recommended versus two subdivisions when the acceptable error is 10 percent. The maximum stress is equal to 21.111 units.

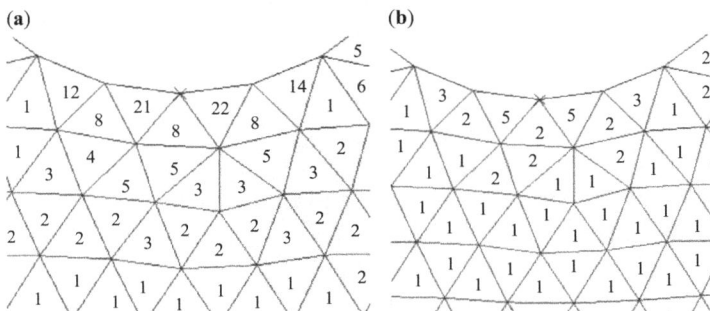

Figure 8.11. Error estimates and refinement guides for σ_x, nominal mesh size = 0.1, acceptable error = 5%: (a) error estimate and (b) refinement guide.

(a)   (b)

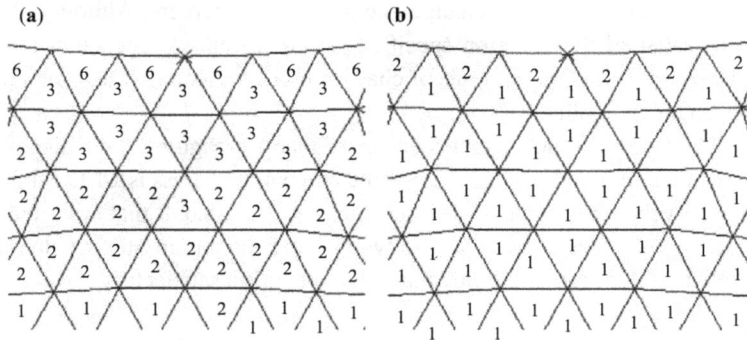

Figure 8.12. Error estimates and refinement guides for σ_x, nominal mesh size = 0.020, acceptable error = 5% (a) error estimate and (b) refinement guide.

When the mesh refinement value of five is applied, the mesh size that is estimated to produce the acceptable level of error of 5 percent or less is equal to h0 = 0.10/5 or h0 = 0.020. When the finite element model with this mesh size is formed and solved, the resulting the error estimates and refinement guides are presented in Figure 8.12.

The maximum stress for this case is 23.968 units. This compares to the maximum stress for the initial mesh of 21.111. The difference between these two values of maximum stress is 11.91 percent. It should be noted that the maximum stress for the case shown in Figure 8.8 where h0 = 0.025 is equal to 24.121 units.

One might anticipate that the finer mesh, that is, h0 = 0.020, should give a more accurate result than the larger mesh, that is, h0 = 0.025. However, this need not be the case because this subdivision is not a *child mesh* of the previous model. A child mesh is a mesh that contains all of the nodes that are contained in the previous mesh in addition to other nodes. This means that the child mesh will represent the problem at least as well as the previous mesh. Therefore, if a mesh is not a child mesh, the result may not be as good as a coarser mesh that better models the exact solution.

When the mesh is refined as suggested, the nominal mesh size is h0 = 0.020/2 = 0.010. When the problem is solved with this mesh, the results are presented in Figure 8.13.

As can be seen, the estimated errors in Figure 8.13a are all below the acceptable limit of 5 percent. As a result, no refinement is recommended in Figure 8.13b. The maximum stress is equal to 24.091 units. This is equal to a difference of 0.50 percent. Thus, we can assume that this is a converged result that can be used in the design process.

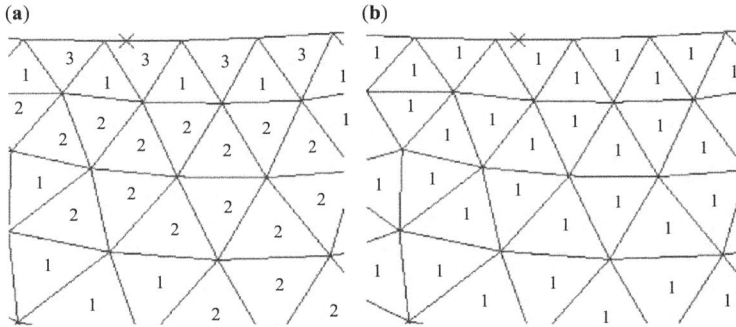

Figure 8.13. Error estimates and refinement guides for σ_x, nominal mesh size = 0.01, acceptable error = 5%: (a) error estimate and (b) refinement guide.

8.9 SUMMARY AND CONCLUSION

The refinement guide presented and demonstrated in this chapter successfully improved the representation of the critical variable in the Kirsch problem. In these examples, the refinement guide focused on improving the representation of the normal stress in the x direction.

The demonstrations presented here are not comprehensive since they focused on a single problem and a single variable. Furthermore, the mesh refinement was not focused only on elements with high levels of estimated error as is the case in the adaptive refinement process. The whole mesh was refined so that a more complex mesh generation program did not have to be introduced.

However, these examples demonstrated the essential features of the refinement guide that was developed here. The basic concepts developed and demonstrated here can be extended to the standard approach to adaptive refinement. This is the case because the error estimator and refinement guide developed here depend on the Taylor series basis of finite element representations.

An individual finite element can only represent a limited number of Taylor series terms. Errors are produced because an element cannot represent the higher-order Taylor series terms that exist in the exact solution. The refinement guide developed here assumes that the majority of the error is due to the inability of an element to represent the first Taylor series terms that the element cannot represent.

In the case of three-node constant strain elements, an element can only represent rigid body motions and constant values for the three strain components. Consequently, the majority of the error in a three-node element is assumed to be due to its inability to represent linear strains. Thus, the refinements estimated for three-node elements vary linearly with the level of error estimates.

On the other hand, a linear strain six-node element can represent linear strain variations, as well as rigid body motions and constant strains. Consequently, the majority of the error in a six-node element is assumed to be due to its inability to represent strains that vary quadratically. Thus, the refinements estimated for six-node elements vary with the square of the level of the error estimates.

As mentioned in the text, a refinement guide that focuses exclusively on the estimate of the first Taylor series term that a given type of element cannot represent is developed in Dow (2012).

8.10   EXERCISES

1. Form a finite difference model for a rectangular plate with zero displacements on the boundary. Load the plate with a point load somewhere near the center. Use nine-node central difference templates to form the model. Find the displacements of the nodes and the strains at each node. Hint: Similar problems are solved in Dow (1999) of Chapter 7.

CHAPTER 9

SUMMARY

9.1 INTRODUCTION

The improvements to the finite element method presented here are derived from a fresh look at the displacement interpolation functions that form the basis of the finite element method. These improvements include the development of: (1) a simple and direct way to evaluate and formulate finite element stiffness matrices, (2) a theoretical basis for creating error estimators, (3) a simple and effective elemental error estimator, and (4) a refinement guide that produces rapidly converging meshes.

9.2 REINTERPRETED INTERPOLATION POLYNOMIALS

The standard form of the displacement interpolation polynomials used in the finite element method has arbitrary coefficients. Examples of these coefficients are contained in the following interpolation polynomials for a three-node element:

$$u(x,y) = a_1 + a_2 x + a_3 y$$
$$v(x,y) = b_1 + b_2 x + b_3 y$$

(9.1)

The salient feature of these coefficients is that they have no specific meaning. In other words, these coefficients can be interpreted as temperatures, displacements, or velocities, depending on the application. Without further analysis, it is impossible to relate these coefficients to the concepts of solid mechanics.

The first step in this reinterpretation of the interpolation polynomials is to recognize that they are truncated Taylor series expansions. They have the following form when they are expressed as Taylor series expansions:

$$u(x,y) = (u)_0 + (\partial u/\partial x)_0 x + (\partial u/\partial y)_0 y$$
$$v(x,y) = (v)_0 + (\partial v/\partial x)_0 x + (\partial v/\partial y)_0 y$$

(9.2)

As can be seen, the coefficients in Equation 9.2 are significantly different than those in Equation 9.1 in that they have physical meaning. These coefficients are functions of the displacements u and v in the x and y directions, respectively. In this form, these coefficients are indirectly related to solid mechanics because displacements are the dependent variables in solid mechanics problems.

However, the coefficients of Equation 9.2 can be specialized so that they directly relate to solid mechanics problems. This is accomplished by expressing the coefficients in terms of rigid body motions and strain quantities. When the definitions of these quantities are introduced into Equation 9.2, we have the following:

$$u(x,y) = (u_{rb})_0 + (\varepsilon_x)_0 x + (\gamma_{xy}/2 - r_{rb})_0 y$$
$$v(x,y) = (v_{rb})_0 + (\gamma_{xy}/2 + r_{rb})_0 x + (\varepsilon_y)_0 y$$

(9.3)

When Equation 9.3 is visually inspected, the six physically interpretable coefficients identify the modeling capabilities of a three-node finite element. The element can represent the three rigid body motions, $(u_{rb})_0$, $(v_{rb})_0$, and $(r_{rb})_0$, and the three constant values of the strain components, $(\varepsilon_x)_0$, $(\varepsilon_y)_0$, and $(\gamma_{xy})_0$.

The first hint of the power of this notation becomes evident when the strain representations for an element are formed from the displacement interpolation functions. When the definitions of strains given in section 3.6 are applied to Equation 9.3, we have the following:

$$\varepsilon_x(x,y) = (\partial u/\partial x) = (\varepsilon_x)_0$$
$$\varepsilon_y(x,y) = (\partial v/\partial y) = (\varepsilon_y)_0$$
$$\gamma_{xy}(x,y) = (\partial v/\partial x + \partial u/\partial y) = (\gamma_{xy})_0$$

(9.4)

Visual inspection of Equation 9.4 shows that a three-node triangle can represent the constant terms of the three strain components. This representation contains no modeling errors. Its only deficiency is that these truncated strain models can only represent constant strains.

When the strain models for the four-node element are formed and visually inspected, Equation 3.8 shows that this element contains several strain modeling errors. In the case of a six-node element, Equation 3.9 shows that the element contains no modeling errors and can represent constant and linearly varying strains. These strain modeling characteristics could not be seen by visual inspect if it were not for the physically interpretable notation.

When the actual strain distribution is too complex to be captured by the modeling capabilities of the finite elements, the result will contain errors. These errors are seen as interelement jumps in the strains. These modeling errors are called discretization errors. They exist because a continuous problem with an infinite number of degrees of freedom has been replaced by a discrete representation with a finite number of degrees of freedom.

9.3   IMPROVEMENTS IN ELEMENT STIFFNESS MATRIX FORMULATION

The introduction of physical meaning into the interpolation polynomials improves the formulation of finite element stiffness matrices in two significant ways. First, the clear identification of the rigid body modes reduces the number of integrals that must be evaluated. This reduction occurs because any integral that contains a rigid body term is equal to zero. This is due to the fact that its contribution to the strain energy is equal to zero. As a result, these terms do not have to be integrated because their value is known by visual inspection.

In addition, the notation simplifies the integrals that must be integrated. As a result of the existence of fewer and simpler integrals, the integrals can be efficiently evaluated exactly. This eliminates the need to use an approximate integration scheme for evaluating these integrals. This, in turn, simplifies the element formulation process.

However, the most significant contribution of this physically based notation is the elimination of strain modeling errors in elements with non-standard shapes, that is, elements with curved edges or elements that are not parallelograms. The very fact that these strain modeling errors exist in the standard isoparametric formulation procedure, albeit, only in certain elements, would seem to render this formulation procedure obsolete.

It should be remembered that the isoparametric formulation procedure was developed when computers possessed only a tiny fraction of the capabilities of today's machines. Consequently, it can be concluded that the isoparametric approach continues to be taught and used because of inertia and the fact that it is embedded in existing code.

9.4 POINTWISE ERROR ESTIMATORS

The error estimators developed in Chapter 7 have a higher resolution than most other error estimators because they evaluate the errors at individual points. The resolution is higher because the highest errors are not submerged in the average error over the domain of the whole element as is the case in error estimators based on strain energy.

The error estimators are put on a solid theoretical foundation because they essentially compare the finite element solution to a finite difference solution. As discussed in Chapter 7, the two approximate solution techniques must converge to the exact solution of the problem. Since the two methods use significantly different approaches to form their approximations, any differences in the approximate solutions are due to deficiencies in the model.

The previous paragraph implies that the error estimates are found by comparing a finite element result with an actual finite difference solution to the problem being solved. However, this is shown to be unnecessary. An approximation of a finite difference solution can be extracted from averaging the nodal quantities in the finite element solution. In other words, an approximation of a finite difference solution can be extracted from the finite element solution in order to accurately estimate the errors in a finite element result.

9.5 AN OVERVIEW OF REFINEMENT GUIDES

The type of refinement guide developed and demonstrated in Chapter 8 essentially compares an estimate of the exact solution that underlies an individual element with the modeling capability of the element. Both the estimate of the underlying solution and the modeling capability of the element are found using physically interpretable notation. As a result, the refinement guide has a solid theoretical basis.

In the simple version of the refinement guide demonstrated in Chapter 8, the estimated error in an element is compared to the level of acceptable error specified in the analysis. This comparison estimates the number of subdivisions that must be given to elements with high levels of error in order to achieve the specified level of error. This error estimator leads to the rapid convergence of the finite element result.

The simple refinement guide demonstrated in Chapter 8 is applied to models formed with constant strain triangles. It has not been tested on higher-order elements. However, another refinement guide that explicitly compares an estimate of the Taylor series expansion of the underlying exact solution to the highest modeling capability of the individual high-error elements is discussed in Chapter 8 (Dow 2012).

If every element in a finite element model satisfies the specified level of error, it does not mean that the solution has converged to the exact solution. It only means that the specified level of error in a stress or strain quantity has been achieved.

As discussed in Chapter 8, it is suggested that the change in the maximum stresses or strains between the iterations of the adaptive refinement procedure be computed. If the change exceeds the acceptable level of error, it is suggested that another improvement to the finite element model be made for safety's sake. This is to insure that the errors in the surrounding elements do not overly affect the critical quantity.

9.6   RESEARCH OPPORTUNITIES

As has been stated, the objective of this book is to provide a clear and easy-to-read overview of adaptive refinement for undergraduates so that the finite element method is readily accessible early in their education. This is deemed beneficial so that less time can be spent learning an analytic tool and more time can be spent using it. However, if students learn how a tool works and how it is constructed well enough that they can use it with confidence, they also can improve it if they so choose.

One of the questions I always put to my students is the following. By now you know that it is wise to presume that everyone working in computational mechanics is intelligent and capable. So why have you and I been able to attack problems that have not been solved before? The correct answer, as I perceive it, is that we have tools available to us that were not previously available.

As noted earlier, all of the developments presented here derive from the use of the physically interpretable notation. Since the transparent nature of this notation provides insights into the finite element method and the finite difference method that have not been previously available, opportunities arise for improvements in these two powerful methods that are only possible now because of this notion.

Some specific research opportunities are identified in Dow (1999, 2012). The most fruitful area may be in the reintroduction of the finite difference method into solid mechanics problems. This method is used extensively in fluid mechanics problems without the use of strain gradient notation. By looking at some of the fluid mechanics techniques with the *new eyes* provided by this notation, who knows what might result? Remember, when pursuing research, the question is often more important than the answer.

ANNOTATED BIBLIOGRAPHY

Argyris, J.H., and S. Kelsey. 1960. *Energy Theorems and Structural Analysis*. London, UK: Butterworths.

This reference contains a series of articles first published in the early 1950's that were some of the earliest material concerning the finite element method.

Boresi, A.P. 1965. *Elasticity in Engineering Mechanics*, 106–09. Englewood Cliffs, NJ: Prentice Hall.

The analysis of the constitutive relations is more extensive in this book than that which is presented in Boresi (1993).

Boresi, A.P., R.J. Schmidt, and O.M. Sidebottom. 1993. *Advanced Strength of Materials*, 5th ed, 101–03. New York: John Wiley.

The two-dimensional case of a constitutive equation with arbitrary coefficients used in this chapter represents a reduction of the three-dimensional case presented in this reference.

Borg, S.F. 1963. *Matrix-Tensor Methods in Continuum Mechanics*, 119–29. New York: Van Nostrand.

This selection of pages reviews the derivation of the compatibility equations in several books and presents a derivation of his own. One can tell that Prof. Borg is a very forgiving man by one statement in his presentation, "The above described procedure is given by Love, Timoshenko, and *other books which follow these*." The italics are mine. Most of these "other books" copy the above described procedure verbatim or with some changes with no additions or clarifications.

Budynas, R.G. 1999. *Advanced Strength and Applied Stress Analysis*, 2nd ed. New York: WCB/McGraw-Hill.

The solution to the Kirsch problem is presented on pages 235–238, The maximum stress concentration for the problem is presented in graphical form on pages 367 and 873.

Dow, J.O., and D.E. Byrd. July, 1981. "A Scale-Free Refined Element Technique." *AIAA Journal* 19, no. 7, pp. 925–32.

In this paper, the idea of using the independence of the stiffness matrix for a two-dimensional problem from size to form "dictionaries" of universally applicable stress concentrations is presented.

Dow, J.O. 1999. *A Unified Approach to the Finite Element Method and Error Analysis Procedures*. New York: Academic Press.

Part II of this book contains a detailed derivation of the strain gradient notation and references to its earliest development. Part III uses strain gradient notation to reformulate the generation of finite element stiffness matrices.

Part IV uses strain gradient notion to identify a common basis for the finite element and finite difference methods and to extend the capabilities of the finite difference method.

Dow, J.O. 2012. *The Essentials of Finite Element Modeling and Adaptive Refinement for Beginning Analysts to Advanced Researchers in Solid Mechanics*. New York: Momentum Press.

Chapter 3 of this book briefly develops this physically interpretable strain gradient notation.

Doyle, A.C. 1917. *His Last Bow: A Reminiscence of Sherlock Holmes*. New York: P. F. Collier and Sons.

Chapter 6 of this book contains the story entitled, *The Disappearance of Lady Frances Carfax*, from which the quote of Sherlock Holmes was taken.

Field, D. 2000. "Qualitative Measures for Initial Meshes." *International Journal for Numerical Methods in Engineering* 4, pp. 709–12.

This book discusses the effect of element geometry on the accuracy of finite element results.

Hughes, T.R.J. 2000. *The Finite Element Method*, 112. Mineola, NY: Dover.

This book includes references to the origin of the isoparametric method.

Lanczos, C. 1966. *The Variational Principles of Mechanics*. Toronto: University of Toronto Press.

McGuire, W., R.H. Gallagher, and R.D. Ziemian. 2000. "Formulation of the Global Analysis Equations." In *Matrix Structural Analysis*, 2nd ed. New York: John Wiley.

Chapter 3—Formation of the Global Analysis Equations presents several examples of assembling the global stiffness matrices for trusses using the implicit approach discussed here in Section 2.4.

Persson, P.-O., and G. Strang. June, 2004. "A Simple Mesh Generator in MATLAB." *SIAM Review* 46, no. 2, pp. 329–45.

This is the source of the mesh refinement program used in this book.

Tiag, I.C. 1961. Structural Analysis by the Matrix Displacement Method. English Electric Aviation Report No. S017.

This book provides background on the development of the isoparametric element formulation process.

Turner, M.J., R.W. Clough, H.C. Martin, and L.J. Topp. September, 1956. "Stiffness and Deflection Analysis of Complex Structures." *Journal of Aeronautical Sciences* 23, no. 9, pp. 805–23.

This is one of the first, if not *the* first, paper to develop finite elements.

Zienkiewicz, O.C. 1998. "As I Remember." Acceptence speech for the Timoshenko Medal, University of Wales, Swansea.

Zienkiewicz originally worked with finite difference solutions of differential equations but ended up writing the first books on finite element because the method was more intuitive.

INDEX

OTHER TITLES IN OUR SOLID MECHANICS COLLECTION

Continuum Mechanics: Basic Principles of Vectors, Tensors, and Deformation
by Tariq A. Khraishi and Yu-Lin Shen

Continuum Mechanics: Constitutive Equations and Applications
by Tariq A. Khraishi and Yu-Lin Shen

Momentum Press is one of the leading book publishers in the field of engineering, mathematics, health, and applied sciences. Momentum Press offers over 30 collections, including Aerospace, Biomedical, Civil, Environmental, Nanomaterials, Geotechnical, and many others.

Momentum Press is actively seeking collection editors as well as authors. For more information about becoming an MP author or collection editor, please visit http://www.momentumpress.net/contact

Announcing Digital Content Crafted by Librarians

Momentum Press offers digital content as authoritative treatments of advanced engineering topics by leaders in their field. Hosted on ebrary, MP provides practitioners, researchers, faculty, and students in engineering, science, and industry with innovative electronic content in sensors and controls engineering, advanced energy engineering, manufacturing, and materials science.

Momentum Press offers library-friendly terms:

- perpetual access for a one-time fee
- no subscriptions or access fees required
- unlimited concurrent usage permitted
- downloadable PDFs provided
- free MARC records included
- free trials

The **Momentum Press** digital library is very affordable, with no obligation to buy in future years.

For more information, please visit **www.momentumpress.net/library** or to set up a trial in the US, please contact **mpsales@globalepress.com.**